# A STORY OF LIGHT

A Short Introduction to Quantum Field Theory of Quarks and Leptons

# A STORY OF LIGHT

## A Short Introduction to Quantum Field Theory of Quarks and Leptons

M. Y. HAN
Duke University, USA

NEW JERSEY • LONDON • SINGAPORE • BEIJING • SHANGHAI • HONG KONG • TAIPEI • CHENNAI

*Published by*

World Scientific Publishing Co. Pte. Ltd.
5 Toh Tuck Link, Singapore 596224
*USA office:* 27 Warren Street, Suite 401-402, Hackensack, NJ 07601
*UK office:* 57 Shelton Street, Covent Garden, London WC2H 9HE

**British Library Cataloguing-in-Publication Data**
A catalogue record for this book is available from the British Library.

**A STORY OF LIGHT**
**An Introduction to Quantum Field Theory of Quarks and Leptons**



ISBN-13 978-981-256-034-6
ISBN-10 981-256-034-3

Typeset by Stallion Press
Email: enquiries@stallionpress.com

Printed in Singapore

For Eema, Grace and Leilani

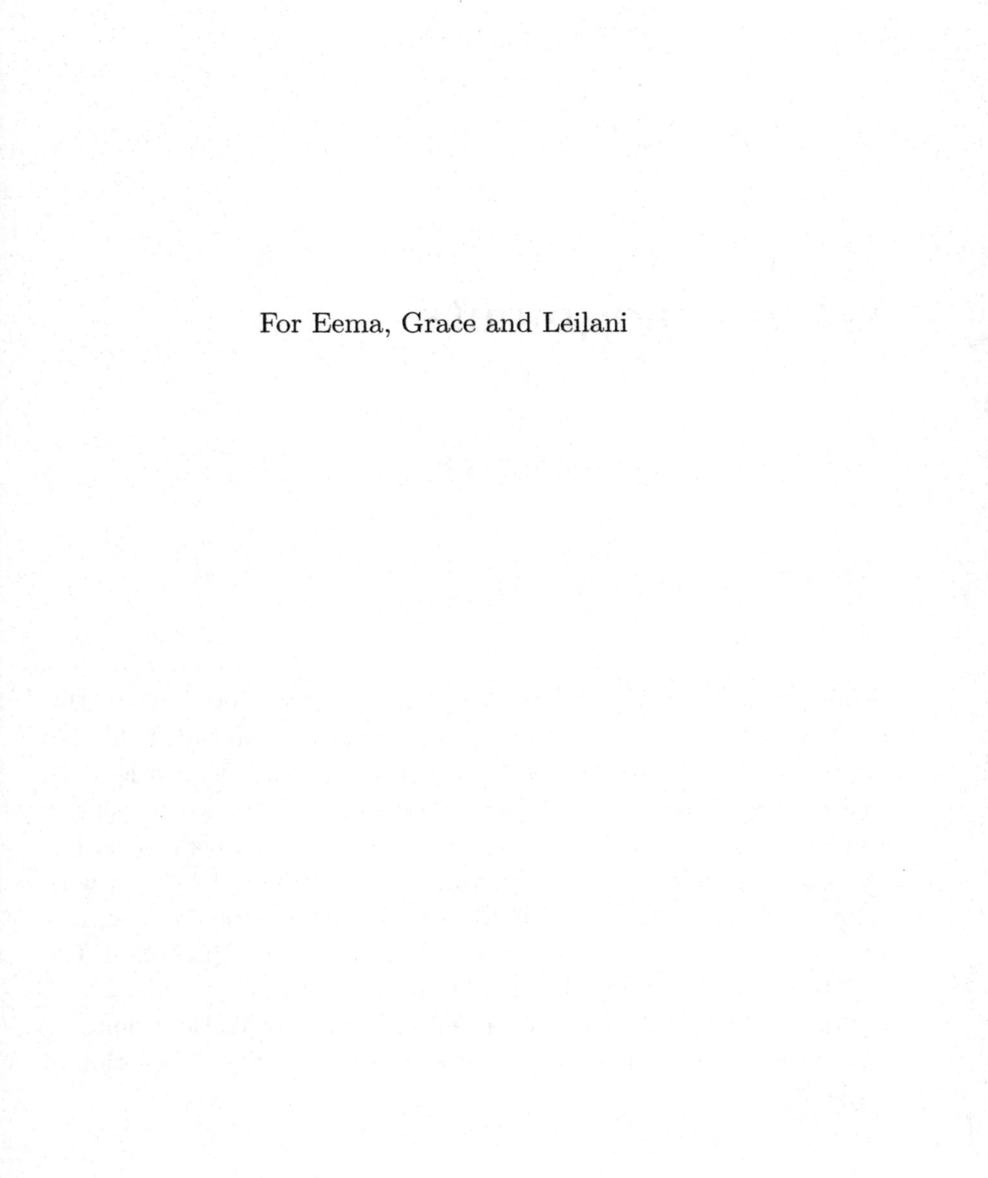

# Acknowledgments

I would like to thank my students who insisted that I write this book after my lectures on the developments in quantum field theory. I would also like to thank Dr. Jaebeom Yoo, a postdoctoral research associate, and Mr. Chang-Won Lee, a graduate student in the Physics Department of Duke University, for valuable discussions and technical help in the preparation of the manuscript. As with my previous book, *Quarks and Gluons*, the constant encouragement from Dr. K.K. Phua, Chairman of World Scientific Publishing Co. is gratefully acknowledged. Thanks are also due to the dedicated help of Ms. Lakshmi Narayan, a Senior Editor of World Scientific, who provided steady and patient guidance toward the completion of this book.

# Contents

# Prologue

The relativistic quantum field theory, or quantum field theory (QFT) for short, is the theoretical edifice of the standard model of elementary particle physics. One might go so far as to say that the standard model *is* the quantum field theory. Having said that as the opening statement of this book, we must be mindful that both quantum field theory and the standard model of elementary particle physics are topics that are not necessarily familiar to many individuals. They are subject areas that are certainly not familiar to those outside the specialty of elementary particle physics, and in some cases not too well grasped even by those in the specialty.

The Standard Model of elementary particle physics is a term that has come into prominence as it became the paradigm of particle physics for the last three decades. In brief, the standard model aims to understand and explain three of the four fundamental forces — the electromagnetic, strong nuclear and weak nuclear — that define the dynamics of the basic constituents of all known matter in the universe.[1] As such, it consists of two interrelated parts: the part

[1]The fourth force of nature, gravity, does not come into play in the scale of the mass of elementary particles and is not included in the standard model. Attempts

that deals with the question of what are the basic building blocks of matter and the second part concerned with the question of what is the theoretical framework for describing the interactions among these fundamental constituents of matter.

A century after the original discovery of quantum of light by Max Planck in 1900 and its subsequent metamorphosis into photon, the zero-mass particle of light, by Albert Einstein in 1905, we have come to identify the basic constituents of matter to be quarks and leptons — the up, down, strange, charm, top and bottom, for quarks, and the electron, muon, tauon, electron-type neutrino, muon-type neutrino, and tauon-type neutrino, for leptons. The three forces are understood as the exchange of "quanta" of each force — photons for the electromagnetic force, weak bosons for the weak nuclear force, and gluons for the strong nuclear force. These particles, some old, such as photons and electrons and some relatively new, such as the top and bottom quarks or the tauons and their associated neutrinos, represent our latest understanding of what are the basic constituents of known matter in the universe.

There are scores of books available which discuss the basic particles of matter, at every level of expertise. For a general readership, we can mention two books that contain no or very little mathematics, *Quarks and Gluons* by myself and *Facts and Mysteries in Elementary Particle Physics* by Martinus Veltman.[2]

The theoretical framework for the three forces or interactions is quantum field theory, that is, the relativistic quantum field theory. Each force has its own form, and again, some old and some new. Quantum electrodynamics, QED for short, was fully developed by the end of the 1940s and is the oldest — and more significantly, the only truly successful quantum field theory to date — of the family. Quantum chromodynamics, QCD, is the framework for the strong nuclear force that is mediated by exchanges of gluons. It was initiated

to merge gravity with the standard model have spawned such ideas as the grand unified theory, supersymmetry, and supersting, the so-called theory of everything. These topics are not discussed in this book.

[2] *Quarks and Gluons* by M. Y. Han, World Scientific (1999); *Facts and Mysteries in Elementary Particle Physics* by Martinus Veltman, World Scientific (2003).

in the 1960s and has been continually developed since, but it is far from becoming a completely successful quantum field theory yet. The theory for the weak nuclear force, in its modern form, was also started in the 1960s, and in the 1970s and 1980s, it was merged with quantum electrodynamics to form a unified quantum field theory in which the two forces — the electromagnetic and weak nuclear — were "unified" into a single force referred to as the electroweak force. Often this new unified theory is referred to as the quantum flavor dynamics, QFD. Thus, the quantum field theory of the standard model consists of two independent components — quantum chromodynamics and quantum flavor dynamics, the latter subsuming quantum electrodynamics.

Despite the abundant availability of books, at all levels, on basic building blocks of matter, when it comes to the subject of relativistic quantum field theory, while there are several excellent textbooks at the graduate level, few resources are available at an undergraduate level. The reason for this paucity is not difficult to understand. The subject of quantum field theory is a rather difficult one even for graduate students in physics. Unless a graduate student is interested in specializing into elementary particle physics, in fact, most graduate students are not required to take a course in quantum field theory. It is definitely a highly specialized course. Quantum field theory thus remains, while a familiar term, a distant topic. Many have not had the opportunity to grasp what the subject is all about, and for those with some rudimentary knowledge of physics at an undergraduate level beyond the general physics, the subject lies well beyond their reach.

The main purpose of this book is to try to fill this gap by bringing out the conceptual understanding of the relativistic quantum field theory, with minimum of mathematical complexities. This book is not at all intended to be a graduate level textbook, but represents my attempt to discuss the essential aspects of quantum field theory requiring only some rudimentary knowledge of the Lagrangian and Hamiltonian formulation of Newtonian mechanics, special theory of relativity and quantum mechanics.

There is another theme in this book and it is this. Throughout the course of development of quantum field theory, from the original quantum electrodynamics in which the Planck–Einstein photon is deemed as the natural consequence of field quantization to the present-day development of the gauge field theory for quarks and leptons, the theories of electromagnetic field have been — and continue to be — a consistently useful model for other forces to emulate. In this process of emulating theories of electromagnetic field, the concept of particles and fields would go through three distinct phases of evolution: separate and distinct concepts in classical physics, the particle-wave duality in quantum mechanics, and finally, particles as the quanta of quantized field in quantum field theory. As we elaborate on this three-stage evolution, we will see that the photon has been — and continues to be — the guiding light for the entire field of relativistic quantum field theory, the theoretical edifice of the standard model of elementary particle physics.

# 1

# Particles and Fields I: Dichotomy

One may have wondered when first learning Newtonian mechanics, also called the classical mechanics, why the concept of a field, the force field of gravity in this case, is hardly mentioned. One usually starts out with the description of motion under constant acceleration — the downward pull of gravity with the value of $9.81\,\mathrm{m/s^2}$. Even when the universal law of gravity is discussed, for example, to explain the Kepler's laws, we do not really get into any detailed analyses of the force field of gravity.

In classical mechanics the primary definition of matter is the point mass, and the emphasis is on the laws of motion for point masses under the influence of force. The focus is on the laws of motion rather than the nature of force field, which is not really surprising when we consider the simplicity of the terrestrial gravitational force field — uniform and in one parallel direction, straight down toward the ground. A point mass is an abstraction of matter that carries mass and occupies one position at one moment of time and this notion of a point mass is diagonally opposite from the notion of a field, which, by definition, is an extended concept, spread out over a region of space.

As we proceed from the study of classical mechanics to that of classical electromagnetism, we immediately notice a big change; from day one it is all about fields. First the electric field, then the magnetic field, and then the single combined entity, the electromagnetic field. No sooner than the Coulomb's law is written down, one defines the electric field and its spatial dependence is determined by Gauss' Law. Likewise, Ampere's Law determines the magnetic field and finally the laws of Faraday and Maxwell lead to the spatial as well as temporal dependence of electromagnetic field.

This dichotomy of the concept of point particle and that of field is in fact as old as the history of physics. From the very beginning, back in the $17^{\text{th}}$ century, there were two distinct views of the physical nature of light. Newton advocated the particle picture — the corpuscular theory of light — whereas Christian Huygens advanced the wave theory of light. For some time — for almost a century and half — these two opposing views remained compatible with what was then known about light — refraction, reflection, lenses, etc. Only when in 1801 Thomas Young demonstrated the wave nature of light by the classic double-slit interference experiment, with alternating constructive and destructive interference patterns, the wave theory triumphed over the particle theory of light.

One might have wondered why the notion of field did not play a prominent role in the initial formulation of Newtonian mechanics, especially since both the gravitational force law and the Coulomb's law obey the identical inverse square force law:

$$F = G\frac{m_1 m_2}{r^2} \quad \text{for gravity}$$

and

$$F = k\frac{q_1 q_2}{r^2} \quad \text{for Coulomb's law}$$

where $G$ and $k$ are the respective force constants, $m$ is mass and $q$ is the electric charge.

The disparity is simply a practical matter of scale. At the terrestrial level, in our everyday world, the inverse square law really does not come into play; the curvature of the surface of the earth is approximated by a flat ground and the gravitational force lines directed toward the center of the earth become, in this approximation, parallel lines pointing downward. In this scale of things, the field aspect of gravity is just too simple to be taken into account. There is no need to bring in any analyses of the gravitational field in the flat surface approximation.

On the contrary, with electric and magnetic forces, we notice and measure in the scale of tabletop experiments the spatial and temporal variations of these fields. The gradients, divergences and curls, to use the language of differential vector calculus, of the electric and magnetic fields come into play in the scale of the human-sized world and this is why the study of electromagnetism always starts off with the definition of electric and magnetic fields.

This well-defined dichotomy of particles and fields, diagonally opposite concepts in classical physics, would evolve through many twists and turns in the twentieth century physics of relativity and quantum mechanics, ending up eventually with the primacy of the concept of field over that of particle in the framework of quantum field theory.

The process of evolution of the concepts of particles and fields have taken a quite disparate path. The Newtonian mechanics has evolved through several steps, some quite drastic. First, there was the Lagrangian and Hamiltonian formulation of mechanics. One of the most important outcomes of this formalism is the definition of what is called the canonically conjugate momentum and this would pave the way for the transition from classical mechanics to quantum mechanics. Quantum field theory could not have developed had it not been the idea of canonically conjugate momentum defined within the Lagrangian and Hamiltonian formalism. As quantum mechanics is merged with special theory of relativity, the culmination of the particle view was reached in the form of relativistic quantum mechanical wave equations, such as the Klein–Gordon and Dirac equations, wherein the wavefunction solutions of these equations provide the relativistic quantum mechanical description of a particle. (More on these equations in later chapters.)

In contradistinction to this development of particle theory, the field view of classical electromagnetism remained almost totally unmodified. The equation of motion for charged particles in an electromagnetic field is naturally accommodated in the Lagrangian and

Hamiltonian formalism. In the Lagrangian formulation of classical mechanics, Maxwell's equations find a natural place by being one of the few examples of what is called the velocity-dependent potentials (more on this in the next chapter). The very definition of the canonically conjugate momentum for charged particles to be the sum of mechanical momentum and the vector potential of the electromagnetic field, discovered back in the $19^{\text{th}}$ century, is in fact the foundation for quantum electrodynamics of the $20^{\text{th}}$ century.

The contrast between the mechanics of particles and the field theory of electromagnetic fields becomes sharper when dealing with the special theory of relativity. The errors of Newtonian mechanics at speeds approaching the speed of light are quite dramatic, and of course, the very foundation of mechanics had to be drastically modified by the relativity of Einstein. Maxwell's equations for the electromagnetic field, on the other hand, required no modifications whatsoever at high speeds; the equations are valid for all ranges of speeds involved, from zero to all the way up to the speed of light. At first, this may strike as quite surprising, but the fact of the matter is that Maxwell's equations lead directly to the wave equations for propagating electromagnetic radiation — light itself. Maxwell's theory of the electromagnetic field is already fully relativistic and hence need no modifications at all.

The development of relativistic quantum mechanics demonstrates quite dramatically the primacy of the classical field concept over that of particles. To cite an important example, in relativistic quantum mechanics, the first and foremost wave equation obeyed by particles of any spin, both fermions of half-integer spin and bosons of integer spin, is the Klein–Gordon equation. Fermions must also satisfy the Dirac equation in addition to the Klein–Gordon equation (more on this in later chapters).

For a vector field $\phi_\mu(x)$ $[\mu = 0, 1, 2, 3]$ for spin one particles with mass $m$, the Klein–Gordon equation is[1]

$$(\partial_\lambda \partial^\lambda + m^2)\phi_\mu(x) = 0$$

[1]Notations and the natural unit system are given in Appendices 1 and 2.

where

$$\partial_\lambda \partial^\lambda = \frac{\partial^2}{\partial t^2} - \nabla^2.$$

For the special case of mass zero particles, of spin one, the Klein–Gordon equation reduces to

$$\partial_\lambda \partial^\lambda \phi_\mu(x) = 0.$$

The classical wave equation for the electromagnetic four-vector potential $A_\mu(x)$, on the other hand, in the source-free region is

$$\partial_\lambda \partial^\lambda A_\mu(x) = 0.$$

An equation for a zero-mass particle of spin one (photon) in relativistic quantum mechanics turns out to be none other than the classical wave equation for the electromagnetic field of the $19^{\text{th}}$ century that predates both relativity and quantum physics!

# 2

# Lagrangian and Hamiltonian Dynamics

Lagrange's equations were formulated by the 18$^{th}$ century mathematician Joseph Louis Lagrange (1736–1813) in his book *Mathematique Analytique* published in 1788. In its original form Lagrange's equations made it possible to set up Newton's equations of motion, $F = dp/dt$, easily in terms of any set of generalized coordinates, that is, any set of variables capable of specifying the positions of all particles in the system. The generalized coordinates subsume the rectangular Cartesian coordinates, of course, but also include angular coordinates such as those in the plane polar or spherical polar coordinates. The generalized coordinates also allow us to deal easily with constraints of motion, such as a ball constrained to move always in contact with the interior surface of a hemisphere; the forces of constraints do not enter into the description of dynamics. As originally proposed, the Lagrange's equations provided a convenient way of implementing Newton's equations of motion.

Lagrange's equations became much more than just a powerful addition to the mathematical technique of mechanics when about 50 years later, in 1834, they became an integral part of Hamilton's principle of least action. Hamilton's principle represents the mechanical form of the calculus of variations that covers wide-ranging fields of physics. Lagrangian and Hamiltonian formulation of mechanics that established the basic pair of dynamical variables — position and momentum — is the precursor to the development of quantum mechanics and when it comes to the development of quantum field theory Lagrangian equations play an absolutely essential role.[1]

For a thorough discourse on the principle of least action in general, and the Hamilton's principle in particular, we will refer readers to many other excellent books on the subject. For our purpose we will focus on specific portions of the Lagrangian and Hamiltonian dynamics that describe the charged particles under the influence of an electromagnetic field. Not always fully appreciated, the Lagrangian and Hamiltonian descriptions of the electromagnetic interaction of the charged particles provide the foundation for quantum electrodynamics, and by extension, for the formulation of the quantum field theories of nuclear forces. The very origin of the field theoretical treatment of electromagnetic interaction traces its root to the classical Lagrangian and Hamiltonian dynamics.

The simplest way to show the equivalence of Lagrange's and Newton's equations is to use the rectangular coordinates, say, $x_i$ ($i = 1, 2, 3$ for more conventional $x, y, z$). Using the notation $\dot{p} = dp/dt$ and $\dot{x} = dx/dt$, Newton's equations are

$$F_i = \dot{p}_i$$
$$p_i = m\dot{x}_i = \frac{\partial}{\partial \dot{x}_i}\left(\frac{1}{2}m\dot{x}_j^2\right) = \frac{\partial T}{\partial \dot{x}_i}$$

where $T$ is the kinetic energy.

For a conservative system

$$F_i = -\frac{\partial V}{\partial x_i}$$

[1]Many excellent standard textbooks on classical mechanics include rich discussions on these subjects — Hamilton's principle, Lagrange's equations, and the calculus of variations. At the graduate level, the *de facto* standard on the subject is *Classical Mechanics* by Herbert Goldstein, Second edition, Addison-Wesley. At an undergraduate level, see, for example, *Classical Dynamics* by Jerry B. Marion, Second edition, Academic Press.

and Newton's equations are transcribed as

$$-\frac{\partial V}{\partial x_i} = \frac{d}{dt}\frac{\partial T}{\partial \dot{x}_i}.$$

In rectangular coordinates (and only in rectangular coordinates)

$$\frac{\partial T}{\partial x_i} = 0$$

and — this is an important point — for a conservative system

$$\frac{\partial V}{\partial \dot{x}_i} = 0.$$

Newton's equations can then be written as

$$\frac{\partial T}{\partial x_i} - \frac{\partial V}{\partial x_i} = \frac{d}{dt}\left(\frac{\partial T}{\partial \dot{x}_i} - \frac{\partial V}{\partial \dot{x}_i}\right)$$

which is Lagrange's equations, usually expressed as

$$\frac{d}{dt}\frac{\partial L}{\partial \dot{x}_i} - \frac{\partial L}{\partial x_i} = 0$$

where $L = T - V$ is the all-important Lagrangian function. The momentum $p$ can be defined in terms of the Lagrangian function as

$$p_i = \frac{\partial L}{\partial \dot{x}_i}.$$

In terms of the generalized coordinates, denoted by $q_i$, that involve angular coordinates in addition to rectangular coordinates, the derivation of Lagrange's equations is slightly more involved. The terms $\partial T/\partial q_i$ are not zero, as in the case of rectangular coordinates, but are fictitious forces that appear because of the curvature of generalized coordinates. For example, in plane polar coordinates, where $T = (m/2)(\dot{r}^2 + r^2\dot{\theta}^2)$, we have $\partial T/\partial r = mr\dot{\theta}^2$, the centrifugal force. Lagrange's equation in terms of generalized coordinates remain in the same form, that is,[2]

$$\frac{d}{dt}\frac{\partial L}{\partial \dot{q}_i} - \frac{\partial L}{\partial q_i} = 0$$

with $L = T - V$ and momentum $p$ is defined by

$$p_i = \frac{\partial L}{\partial \dot{q}_i}.$$

This definition of momentum $p$ in terms of the Lagrangian represents a major extension of the original definition by Newton. In rectangular coordinates, it reduces to its original form, of course, but for those generalized coordinates corresponding to angles the new momentum corresponds to the angular momentum. In plane polar coordinates where $T = (m/2)(\dot{r}^2 + r^2\dot{\theta}^2)$, and $\partial V/\partial\dot{\theta} = 0$,

$$p_\theta = \frac{\partial L}{\partial \dot{\theta}} = \frac{\partial T}{\partial \dot{\theta}} = mr^2\dot{\theta}$$

which is the angular momentum corresponding to the angular coordinate.

This new definition of momentum is technically called the canonically conjugate momentum, that is, $p_i$ being conjugate to the generalized coordinate $q_i$, and this pairing of $(q_i, p_i)$ forms the very basis of the development of quantum mechanics and, by extension, the quantum field theory. Having thus become the standard basic dynamical variables, they are simply referred to as coordinates (dropping "generalized") and momenta (dropping "canonically conjugate").

This new definition of the canonically conjugate momentum, or simply momentum, has far-reaching consequences when the Lagrangian formulation is adopted to the case of charged particles interacting with the electromagnetic field. Often mentioned as a supplement within the framework of classical mechanics, this casting of Maxwell's equations into the framework of Lagrangian formulation

[2]Usually, Lagrange's equations are first derived from other physical principles — D'Alembert's principle or Hamilton's principle — and their equivalence to Newton's equations is shown to follow from the former. Here we follow the derivation as given in *Mechanics* by J.C. Slater and N.H. Frank, McGraw-Hill, which starts from Newton's equation, and then derive Lagrange's equations.

leads to non-mechanical extension of momentum and, as we will follow through in later chapters, provides the very foundation for the development of quantum electrodynamics.

Almost all forces we consider in mechanics are conservative forces, those that are functions only of positions, and certainly not functions of velocities, that is, $\partial V/\partial \dot{q}_i = 0$. There is, however, one very important case of a force that is velocity-dependent, namely, the Lorentz force on charged particles in electric and magnetic fields. In an amazing manner, the velocity-dependent Lorentz force fits perfectly into the Lagrangian formulation.

The Lagrangian equation can be written as

$$\frac{d}{dt}\frac{\partial T}{\partial \dot{q}_i} - \frac{\partial T}{\partial q_i} = -\frac{\partial V}{\partial q_i} + \frac{d}{dt}\frac{\partial V}{\partial \dot{q}_i}.$$

For conservative systems, $\partial V/\partial \dot{q}_i = 0$. For non-conservative systems when forces, and their potentials, are velocity-dependent, it is possible to retain Lagrange's equations provided that the velocity-dependent forces are derivable from velocity-dependent potentials — also called the generalized potentials — in specific form as required by Lagrange's equations, namely, the force is derivable from its potential by the recipe, expressed back in terms of the rectangular coordinates,

$$F_i = \left(-\frac{\partial}{\partial x_i} + \frac{d}{dt}\frac{\partial}{\partial \dot{x}_i}\right) V.$$

It is a rather stringent requirement and it turns out — very fortunate for the development of quantum electrodynamics — that the Lorentz force satisfies such requirement.

Putting $c = \hbar = 1$ in the natural unit system, the Lorentz force on a charge $q$ in electric and magnetic fields, $\mathbf{E}$ and $\mathbf{B}$, is given by

$$\mathbf{F} = q\mathbf{E} + q(\mathbf{v} \times \mathbf{B})$$

where

$$\mathbf{E} = -\nabla\phi - \frac{\partial \mathbf{A}}{\partial t} \quad \text{and} \quad \mathbf{B} = \nabla \times \mathbf{A}$$

and $\phi$ and $\mathbf{A}$ are the scalar and vector potentials, respectively, defining the four-vector potential $A_\mu = (\phi, \mathbf{A})$. After some algebra,[3] the

[3]See Appendix 3.

Lorentz force can be expressed as

$$F_i = -\frac{\partial U}{\partial x_i} + \frac{d}{dt}\frac{\partial U}{\partial \dot{x}_i}$$

where

$$U = q\phi - q\mathbf{A} \cdot \mathbf{v}.$$

The Lagrangian for a charged particle in an electromagnetic field is thus

$$L = T - q\phi + q\mathbf{A} \cdot \mathbf{v}$$

and, as a result, the momentum — the new canonically conjugate momentum — becomes

$$\mathbf{p} = m\mathbf{v} + q\mathbf{A},$$

that is, mechanical Newtonian momentum plus an additional term involving the vector potential.

The Lagrangian formulation of mechanics was then followed by the Hamiltonian formulation based on treating the conjugate pairs of coordinates and momenta on an equal footing. This then led to the Poisson brackets for $q$'s and $p$'s and the Poisson brackets in turn led directly to quantum mechanics when they were replaced by commutators between the conjugate pairs of dynamical variables. For our purpose, we again focus on the motion of charged particles in an electromagnetic field.

In the Hamiltonian formulation, the total energy of a charged particle in an electromagnetic field is given by

$$E = \frac{1}{2m}(p_j - qA_j)(p_j - qA_j) + q\phi.$$

Comparing this expression to the total energy $E$ for a free particle

$$E = \frac{1}{2m}p_j p_j$$

($p_j$ in each expression is the correct momentum for that case, that is, for free particles $m\dot{x}_j = p_j$, but for charged particles in an electromagnetic field $m\dot{x}_j = p_j - qA_j$), we arrive at the all-important substitution rule: the electromagnetic interaction of charged particles is given by replacing

$$E \Rightarrow E - q\phi$$

and

$$\mathbf{p} \Rightarrow \mathbf{p} - q\mathbf{A}.$$

In relativistic notations, this substitution rule becomes a compact expression

$$p^\mu \Rightarrow p^\mu - qA^\mu$$

where

$$p^\mu = (E, \mathbf{p}) \quad \text{and} \quad A^\mu = (\phi, \mathbf{A}).$$

As we shall see later, this substitution rule, obtained when Maxwell's equations for the electromagnetic field are cast into the framework of Lagrangian and Hamiltonian formulation of mechanics, is the very foundation for the development of quantum electrodynamics and, by extension, quantum field theory. It is *that* important. One must also note that whereas the Newtonian dynamics for particles went through modifications and extensions by Lagrange and Hamilton, the equations for the electromagnetic field not only remain unmodified but also, in fact, yielded a hidden treasure of instructions on how to incorporate the electromagnetic interaction.

# 3

# Canonical Quantization

Transition from classical to quantum physics, together with the discovery of relativity of space and time, represents the beginning of an epoch in the history of physics, signaling the birth of modern physics of the $20^{\text{th}}$ century. Quantum physics consists, broadly, of three main theories — non-relativistic quantum mechanics, relativistic quantum mechanics, and the quantum theory of fields. In each case, the principle of quantization itself is the same and it is rooted in the canonical formalism of the Lagrangian and Hamiltonian formulation of classical mechanics. In the Hamiltonian formulation, the coordinates and momenta are accorded an equal status as independent variables to describe a dynamical system, and this is the point of departure for quantum physics.

The two most important quantities in the Hamiltonian formulation is the Hamiltonian function and the Poisson bracket. The Hamiltonian function $H$ — or just Hamiltonian, for short — is defined by

$$H(q, p, t) = \dot{q}_i p_i - L(q, \dot{q}, t)$$

where $L$ is the Lagrangian. In all cases that we consider, $\dot{q}_i p_i$ is equal to twice the kinetic energy, $T$, and with the Lagrangian being equal to

$T - V$, the Hamiltonian corresponds to the total energy of a system, namely,

$$H = T + V.$$

The Poisson bracket of two functions $u$, $v$ that are functions of the canonical variables $q$ and $p$ is defined as

$$\{u, v\} = \frac{\partial u}{\partial q_i}\frac{\partial v}{\partial p_i} - \frac{\partial u}{\partial p_i}\frac{\partial v}{\partial q_i}.$$

When $u$ and $v$ are $q$'s and $p$'s themselves, the resulting Poisson brackets are called the fundamental Poisson brackets and they are:

$$\{q_j, q_k\} = 0,$$
$$\{p_j, p_k\} = 0,$$

and

$$\{q_j, p_k\} = \delta_{jk}.$$

In terms of the Poisson brackets, the equations of motion for any functions of $q$ and $p$ can be expressed in a compact form. For some function $u$ that is a function of the canonical variables and time, we have

$$\frac{du}{dt} = \{u, H\} + \frac{\partial u}{\partial t}$$

where $\{u, H\}$ is the Poisson bracket of $u(q, p, t)$ and the Hamiltonian $H$.

The transition from the Poisson bracket formulation of classical mechanics to the commutation relation version of quantum mechanics is affected by the formal correspondence ($\hbar$ is set equal to 1):

$$\{u, v\} \Rightarrow \frac{1}{i}[u, v]$$

where $[u, v]$ is the commutator defined by $[u, v] = uv - vu$, and on the left $u$, $v$ are classical functions and on the right they are quantum mechanical operators. This transition from functions to operators and from the Poisson brackets to commutators is the very essence of quantization in a nut-shell.

In quantum mechanics, the time dependence of a system can be ascribed to either operators representing observables — momentum, energy, angular momentum and so on — or to wavefunctions representing the quantum state of a system. The former is called the Heisenberg picture and the latter Schrödinger picture (the third option is what is called the interaction picture in which both wavefunctions and operators are functions of time). In the Heisenberg picture, the equations of motion for any observable $U$ is given by an exact counterpart of the classical equation in terms of the Poisson brackets, but with the Poisson bracket replaced by commutator, that is:

$$\frac{dU}{dt} = \frac{1}{i}[U, H] + \frac{\partial U}{\partial t}.$$

In the Schrödinger picture, operators representing observables are built up from those representing (canonically conjugate) momentum and energy by differential operators (expressed in rectangular coordinates),

$$p_j = -i\frac{\partial}{\partial x_j}$$

and since time $t$ and $-H$ are also canonically conjugate to each other

$$E = i\frac{\partial}{\partial t}.$$

The wave equations of quantum mechanics, both non-relativistic and relativistic, are usually expressed in the Schrödinger picture and we have in the case of the non-relativistic quantum mechanics the time-dependent and time-independent Schrödinger's equations,

$$i\frac{\partial\psi(x_i, t)}{\partial t} = E\psi(x_i, t) \text{ (time-dependent)}$$

and from $E = p^2/(2m) + V$

$$\left(-\frac{1}{2m}\nabla^2 + V\right)\phi(x_i) = E\phi(x_i) \text{ (time-independent)}$$

where $\phi(x_i)$ is the space-dependent part of the total wavefunction $\psi(x_i, t)$. There are several wave equations in the relativistic quantum mechanics (Klein–Gordon, Dirac, Proca and other equations), but they must all first and foremost satisfy the relativistic energy–momentum relations

$$E^2 = p^2 + m^2$$

from which we obtain the Klein–Gordon equation

$$\left(\frac{\partial^2}{\partial t^2} - \nabla^2 + m^2\right)\phi(x) = 0$$

or

$$(\partial_\lambda \partial^\lambda + m^2)\phi(x) = 0$$

that was mentioned in Chapter 1. As will be seen later, the quantum field theory is completely cast in the Heisenberg picture wherein the quantum mechanical wavefunctions themselves become operators.

The canonical procedure of quantization, be it non-relativistic quantum mechanics, relativistic quantum mechanics, or relativistic quantum field theory, can thus be compactly summarized as follows.

(i) First, find the Lagrangian function $L$ which yields the correct equations of motion via the Lagrange's equation

$$\frac{d}{dt}\frac{\partial L}{\partial \dot{q}_i} - \frac{\partial L}{\partial q_i} = 0.$$

In the case of mechanical systems, $L = T - V$ and the equations of motion reduce to Newton's equations of motion.

(ii) Define canonically conjugate momentum $p$ with the help of $L$,

$$p_i = \frac{\partial L}{\partial \dot{q}_i}.$$

(iii) Quantization is affected when we impose the basic commutation relations

$$[q_j, q_k] = [p_j, p_k] = 0$$

and

$$[q_j, p_k] = i\delta_{jk}.$$

(iv) The wave equations in quantum mechanics, both non-relativistic and relativistic, are obtained, in the Schrödinger picture, by the operator representation of momentum and energy (expressed in rectangular coordinates) as

$$p_j = -i\frac{\partial}{\partial x_j} \quad \text{and} \quad E = i\frac{\partial}{\partial t}.$$

As will be seen later, in the quantum field theory the very quantum mechanical wavefunctions themselves become operators for generalized coordinates and the corresponding canonically conjugate momenta are defined by the same recipe via the Lagrangian. The Lagrangian function thus is an absolutely essential element in any quantum physics, be it quantum mechanics or quantum field theory.

# 4

# Particles and Fields II: Duality

The departure of quantum mechanics from classical mechanics is quite drastic, rather extreme in contemporary parlance. Ordinary physical quantities are replaced by quantum mechanical operators that do not necessarily commute with each other and the Heisenberg's uncertainty principles between the canonically conjugate pairs of variables, between coordinates and momenta and between time and energy, deny the complete determinability of classical physics.

The most basic and defining characteristic of quantum mechanics — often called the central mystery of quantum mechanics — is the uniquely dual nature of matter called the wave–particle duality. In the microscopic scale of quantum world — of atoms, nuclei and elementary particles — a physical object behaves in such a way that exhibits the properties of both a wave and a particle. Often the wave–particle duality of quantum world is presented as physical objects that are *both* a wave and a particle. To us, with human intuition being nurtured in the macroscopic world, this simplistic picture of wave–particle duality is, of course, completely counterintuitive.

A quantum mechanical object is actually neither a wave in the classical sense nor a particle in the classical sense, but rather it defines a totally new reality, the quantum reality, that in some circumstances exhibits properties much like those of classical particles and in some other circumstances displays properties much like those of classical wave. The new quantum reality can be stated as being "neither a wave nor a particle but is something that can act sometimes much like a wave and at other times much like a particle." The new quantum reality, the wave–particle duality, thus combines the classical dichotomy of particles and fields, waves being specific examples of a field broadly defined as an entity with spatial extension.

Let us briefly recapitulate what is meant by a particle in the classical sense. First, it has mass and occupies one geometric point in space; that is, it has no spatial extension. When it moves, under the influence of a force, it moves from one point at one time to another point at another moment in time. The entire trace of its motion is called its trajectory. Once the initial position and velocity are fixed, Newton's equations of motion determine completely its trajectory. If the laws of motion dictate a particle to pass through a particular position $A$ at some time $t$, the particle will pass through that point. There is no way the particle can be seen to be passing through any other positions at that same time. It will pass through the position $A$ and nowhere else. Furthermore, without any force to alter its course, a particle cannot simple decide to change its direction of motion and can go to other positions. That is a no–no.

Another defining characteristic of a particle in the classical sense is the way in which it impacts, that is, how it interacts with another object. The classical particle interacts with others at a point of collision; some of its energy and momentum are transferred to others at that point of impact. It is the point-to-point transfer of energy and momentum that is the basic dynamical definition of what a particle is in the classical sense.

We can easily contradistinguish the kinematical and dynamical aspects of a classical wave from those of a classical particle. First and foremost, a wave is certainly not something that is defined at a geometrical point. On the contrary, a wave is definitely an extended object — a wave train with certain wavelength and frequency — and furthermore it does not travel along a point-to-point trajectory, but rather propagates in all directions. A sound or light wave propagates from its source in expanding spheres in all directions. In a room whose walls are shaped as the interior surface of a sphere, a wave will hit all points of the wall at the same time.

As a wave also carries its own energy and momentum, the way it propagates in all directions dictates the way it transfers energy and momentum, everywhere in all directions as it comes into contact with other matter. There is no point-to-point transfer as far as the wave is concerned. The contrast between the classical particle and wave could not have been more diagonally opposite, and this is what the new quantum reality called wave–particle duality brings together!

Now, an important caveat is in order about a matter of terminology in quantum physics. The new quantum reality, the wave–particle duality, describes a quantum *thing* that is neither a particle in classical sense nor a wave in classical sense. We can shorten the name to simply *duality*, that is, electrons, protons, neutrons and photons, etc should all be called *duality*, certainly not *particle* nor *wave*. Our reluctance or inability to part with the word "particle" is such that, however, the objects in the quantum world — be they electrons, photons, protons, neutrons, quarks and whatever — are continually referred to as "particles," as in unstable particles, elementary particle physics and so on. What has happened is that the meaning of the word "particle" has gone through a metamorphosis: the word particle when applied to entities in the quantum world actually means duality, the wave–particle duality. Terminologies have gone from classical wave and classical particle to wave–particle duality, or just duality for short, and the "duality" has morphed back into "particle." We will conform to this practice and from this point on in this book, the word "particle" will stand for duality and the word "particle" in classical sense will always be referred to as, "particle in the classical sense." This is a somewhat confusing story of the evolution in the meaning of the word "particle."

The properties of this (new quantum) particle are thus this. When it impacts, that is, interacts with other particles, it behaves much as the way a classical particle does, that is, a transfer of energy and momentum occurs at a point. But it travels in all directions like a classical wave. Since the impact occurs at a point, the question then arises as to what determines the particle to impact at one point at some time and at a different point at another time. In other words, what determines its preference to land at a particular point, and not elsewhere, at one time and land at another point at another time. This is the crux of the matter of quantum mechanics: the particle carries with it the information that determines the probability of its landing at a particular point.

Going back to the example of a room with the walls shaped like the interior walls of a sphere, the (quantum) particle can strike anywhere on the wall, impacting a particular point as if it were a classical particle (making a point mark on the wall). If you repeat the experiment again and again, the particle will land at points (one at a time) all over the wall, but with varying probabilities, at some points more often than at some other points.

The defining properties of the (quantum) particle can thus be summarized as:

(i) It spreads like a classical wave, in all directions.
(ii) However, it impacts like a classical particle.
(iii) It carries with it its own information on the probability of where it is likely to impact.

The next logical question then is what determines its probability. And this is what the equations of quantum mechanics, such as the Schrödinger's equation of non-relativistic quantum mechanics, provide as their solutions, namely, wavefunctions.

This is phase II in the evolution of particles and fields, first the classical dichotomy and now the quantum duality.

# Equations for Duality

The wavefunctions for a particle (in the new sense of wave–particle duality) are to be determined as solutions of quantum mechanical wave equations and these wavefunctions provide information on the probability of the particle impacting at or near a particular position. We have already mentioned the quantum mechanical differential operators corresponding to momentum and energy and by substituting these operator expressions to either the non-relativistic formula for total energy or the relativistic one, we obtain the corresponding equations of quantum mechanics.

## The Schrödinger's Equation

In non-relativistic quantum mechanics, the equation in question is the Schrödinger's equation, which is the central and only wave equation for non-relativistic quantum mechanics. As mentioned in Chapter 3, the time-dependent Schrödinger's equation is

$$i\frac{\partial\psi(x_i,t)}{\partial t} = E\psi(x_i,t).$$

Writing $\psi(x_i, t) = \phi(x_i)T(t)$, the function of time only has solutions in the form of

$$T(t) = \exp(-iEt)$$

where $E$ is the quantized values (eigenvalues) of energy determined from the time-independent Schrödinger's equation obtained from

$$\left(-\frac{1}{2m}\nabla^2 + V\right)\phi(x_i) = E\phi(x_i).$$

The absolute square of the solutions $|\phi(x_i)|^2$, for each allowed values of $E$, is the probability distribution function of finding the particle in question in a state with a particular value of energy, $E$, in a small region between $x$ and $x + dx$. The wavefunctions are referred to as the probability amplitudes, or just amplitudes, and the absolute square of wavefunctions as the probability density, or just probability. This interpretation, the postulate of probabilistic interpretation of wavefunctions, is one of the basic tenets of quantum mechanics and is the "heart and soul" of wave–particle duality.

The Schrödinger's equation and its solutions, however, fall short of accommodating one of the basic attributes of particles, the spin of a particle. Since the spin is an intrinsic property of a particle that is not at all associated with the spatial and temporal coordinates of the particle, it is one of the internal degrees of freedom of a particle — as opposed to the spatial and temporal coordinates being the external degrees of freedom — and the Schrödinger's equation is not set up to deal with any such internal degrees of freedom. The electric charge of a particle is another example of internal degrees of freedom that has nothing to do with the spatial and temporal coordinates.

In the case of electronic orbits of an atom, for example, the solutions $\phi(x_i)$ successfully specify the radii of the orbits (the principal quantum number), the angular momentum of an electron in an orbit (the total angular momentum quantum number), and the tilts of the planes of an orbit (the magnetic quantum number). The complete knowledge of the structure and physical properties of atoms, however, requires specification of the electron spin and Pauli's exclusion principle, without which the physics of atoms, and by extension, all known matter in the universe, would not have been what it is. In this sense, while it is the enormously successful central equation for atomic physics, the Schrödinger's equation falls short of completing the story of atoms. The spin part of information is simply tacked onto the wavefunctions as an add-on, in the case of electrons, by a two-component (for spin-up and spin-down) one-column matrices.

The spin of a particle finds its rightful place only when we proceed to relativistic quantum mechanics. Particles with half-integer spin — generically called the fermions — such as electrons, protons and neutrons that constitute all known matter obey the relativistic wave equation called the Dirac equation (see below), wherein only the total angular momentum defined as the sum of orbital angular momentum and spin is conserved, whereas in the non-relativistic case the conserved quantity is the orbital angular momentum only. The wave equations for relativistic quantum mechanics are obtained from the relativistic energy momentum relation by an operator substitution

$$p^\mu = i\partial^\mu = i\frac{\partial}{\partial x_\mu} = i\left(\frac{\partial}{\partial t}, -\nabla\right)$$

into

$$E^2 - p^2 = m^2 \quad \text{or} \quad p^\mu p_\mu = m^2.$$

### The Klein–Gordon Equation

Particles with spin zero, those with no spin at all, are described by a scalar amplitude, $\phi(x)$, that is invariant under the Lorentz transformation, meaning that the amplitude remains the same as observed in any inertial frame. For brevity, we will use the notation $x$ for space-time coordinate four-vector (Appendix 2, Notations). From $p^\mu p_\mu = m^2$, we have the Klein–Gordon equation

$$(\partial^\mu \partial_\mu + m^2)\phi(x) = 0.$$

The Schrödinger wavefuntion is also a scalar wavefunction; it does not address the spin degrees of freedom. For particles of other values of spin, spin one for vector bosons and spin one-half for fermions, the wavefunctions are not scalars; they are four-vector wavefunctions for spin one vector bosons and four-component spinors for fermions, and each satisfies its own set of equations over and beyond the Klein–Gordon equation. But any relativistic wavefunction, regardless of the spin of the particle, must first and foremost satisfy the Klein–Gordon equation.

### The Dirac Equation

The Dirac equation is the most significant achievement of relativistic quantum mechanics. It successfully incorporated the spin of a particle as the necessary part of the particle's total angular momentum, and it also predicted the existence of antiparticles — positrons, antiprotons, antineutrons, and so forth. Since 99.9% of the known matter in the universe is made up of electrons, protons and neutrons, all of which are spin one-half fermions, the Dirac equation applies to the basic particles that make up all known matter. One can go so far as to claim that the Dirac equation and relativistic quantum mechanics are virtually synonymous.

What originally prompted Dirac to search for and discover the Dirac equation is simple and straightforward enough. The Klein–Gordon equation is a second-order differential equation — second derivatives with respect to both space and time — and as a relativistic equation for single particle, it encounters some difficulties; the nature of second-order differential equations and the probability interpretation of quantum mechanics clash. (We will not discuss these difficulties here, but mention that difficulties do arise for the Klein–Gordon equation as one-particle equation becomes resolved when solutions of the Klein–Gordon equation are treated as quantized fields in quantum field theory.) Rather than a second-order equation, Dirac wanted a first-order linear equation containing only the first derivatives with respect to both space and time, that is, linear with respect to four-vector derivates.

The process of going from a second-order expression to a first-order one is a matter of factorization and let us dwell on this matter here. The simplest algebraic factorization is, of course, the factorization of $x^2 - y^2$:

$$x^2 - y^2 = (x + y)(x - y).$$

Factorization of $x^2 + y^2$, however, cannot be done in terms of real numbers but needs the help of complex numbers:

$$x^2 + y^2 = (x + iy)(x - iy).$$

Factorization of a three-term expression such as $x^2 + y^2 + z^2$ requires much more than just numbers, real or complex; and we must rely on matrices. Consider the three Pauli spin matrices $\sigma$, given in their standard representation as

$$\sigma_x = \begin{pmatrix} 0 & 1 \\ 1 & 0 \end{pmatrix}, \quad \sigma_y = \begin{pmatrix} 0 & -i \\ i & 0 \end{pmatrix}, \quad \sigma_z = \begin{pmatrix} 1 & 0 \\ 0 & -1 \end{pmatrix},$$

satisfying the anticommutation relations

$$\{\sigma_j, \sigma_k\} \equiv \sigma_j \sigma_k + \sigma_k \sigma_j = 2\delta_{jk}.$$

For any two vectors $\mathbf{A}$ and $\mathbf{B}$ that commute with $\sigma$, we have the following identity

$$(\sigma \cdot \mathbf{A})(\sigma \cdot \mathbf{B}) = \mathbf{A} \cdot \mathbf{B} + i\sigma \cdot (\mathbf{A} \times \mathbf{B}).$$

When applied to only one vector, the identity reduces to

$$(\sigma \cdot \mathbf{A})(\sigma \cdot \mathbf{A}) = \mathbf{A} \cdot \mathbf{A}$$

and this allows, in terms of $2 \times 2$ anticommuting matrices, factorization of three-term expressions, such as

$$p_x^2 + p_y^2 + p_z^2 = \mathbf{p} \cdot \mathbf{p} = (\sigma \cdot \mathbf{p})(\sigma \cdot \mathbf{p}).$$

Now we can factorize $p^\mu p_\mu = E^2 - (p_x^2 + p_y^2 + p_z^2)$, a four-term expression is thus

$$\begin{aligned} p^\mu p_\mu &= E^2 - (p_x^2 + p_y^2 + p_z^2) = E^2 - (\sigma \cdot \mathbf{p})(\sigma \cdot \mathbf{p}) \\ &= (\mathrm{E} + \sigma \cdot \mathbf{p})(\mathrm{E} - \sigma \cdot \mathbf{p}). \end{aligned}$$

This has led to the relativistic wave equation for massless fermions in the form of

$$p^\mu p_\mu \varphi_\alpha = (\mathrm{E} + \sigma \cdot \mathbf{p})(\mathrm{E} - \sigma \cdot \mathbf{p})\varphi_\alpha = 0 \quad \text{with } \alpha = 1, 2,$$

where $\varphi_\alpha$ is a two-component wavefunction (since the Pauli matrices are $2 \times 2$ matrices). This used to be the wave equation for two-component zero-mass electron neutrinos (nowadays, the neutrinos are considered to have mass, however minute it may be). Factorization by the use of three $2 \times 2$ matrices renders the amplitude $\varphi_\alpha(x)$ to be a two-component column matrix, called a spinor.

This then brings us to the Dirac equation as a result of factorizing the five-term expression of the Klein–Gordon equation, $p^\mu p_\mu - m^2$. It cannot be brought to a linear equation even with the help of three $2 \times 2$ Pauli matrices. Dirac has shown that factorization is possible but only with the help of four $4 \times 4$ matrices that are built up from the $2 \times 2$ Pauli matrices. Such is the humble beginning of the Dirac equation that came to govern the behavior of all particles of half-integer spin. The five-term expression can be factorized thus

$$\begin{aligned} p^\mu p_\mu - m^2 &= E^2 - (p_x^2 + p_y^2 + p_z^2) - m^2 \\ &= (\gamma^\mu p_\mu + m)(\gamma^\nu p_\nu - m) \end{aligned}$$

where the four $\gamma^\mu$ matrices are required to satisfy the anticommutation relations

$$\gamma^\mu \gamma^\nu + \gamma^\nu \gamma^\mu = 2g^{\mu\nu}$$

and by virtue of which

$$\gamma^\mu p_\mu \gamma^\nu p_\nu = p^\mu p_\mu .$$

It is the four-dimensional analogue of the three-dimensional relations

$$(\sigma \cdot \mathbf{p})(\sigma \cdot \mathbf{p}) = \mathbf{p} \cdot \mathbf{p}.$$

Of the many different matrix representations of four $\gamma$ matrices, the most-often used is where

$$\gamma^0 = \begin{pmatrix} I & 0 \\ 0 & -I \end{pmatrix}, \quad \text{and} \quad \gamma^k = \begin{pmatrix} 0 & \sigma^k \\ -\sigma^k & 0 \end{pmatrix}$$

and the $\sigma$'s are Pauli's spin matrices and $I$ is the $2 \times 2$ unit matrix.

The Dirac equation then becomes, replacing $p_\mu$ by $i\partial_\mu$,

$$(i\gamma^\mu \partial_\mu - m)\psi_\alpha(x) = 0.$$

The Dirac amplitude $\psi_\alpha(x)$ with $\alpha = 1, 2, 3, 4$ is now a four-component spinor which, in the standard representation, consists of positive-energy solutions with spin up and down and negative-energy solutions with spin up and down.

It is clear from the factorization of the second-order relativistic energy momentum relation that each component of the Dirac amplitude must also independently satisfy the Klein–Gordon equation, that is,

$$(\partial^\mu \partial_\mu + m^2)\psi_\alpha(x) = 0 \qquad \text{for each } \alpha = 1, 2, 3, 4.$$

The Dirac equation imposes further conditions over and beyond the Klein–Gordon equation — very stringent interrelations among the components and their first derivatives — among the four components of the solution. This can be seen when the Dirac equation is fully written out in $4 \times 4$ matrix format using an explicit representation of $\gamma$-matrices such as shown above.

The Dirac equation is the centerpiece of relativistic quantum mechanics. All textbooks on the subject devote a substantial amount of the contents to detailing all aspects of this equation — proof of its relativistic covariance, the algebraic properties of Dirac matrices, as $\gamma$ matrices are called, the bilinear covariants built from its four-component solutions, and many others — and, in fact, virtually all textbooks on quantum field theory also include extensive discussions about the equation, before embarking on the subject of field quantization. We will not discuss here the extensive properties of Dirac equation, but suffice it to say that the equation is perhaps the most important one in all of quantum mechanics. It is an absolutely essential tool in elementary particle physics. After all, it is the equation for all particles that constitute the known matter in the universe — all fermions of spin one-half which also includes all leptons, of which the electron is the premier member, and all quarks, out of which such particles as protons and neutrons are made. (More on leptons and quarks in later chapters.)

At this point, let us briefly recap what we traced out in the previous five chapters, including this one. The evolution in our treatment of material particles has come through several phases. The abstract concept of a point mass in Newtonian mechanics remained intact through the development of Lagrangian and Hamiltonian dynamics. When quantum mechanics replaced the classical dynamics of Newton, Lagrange and Hamilton, the concept of particle went through a fundamental revision, from that of a well-defined classical point mass to one of quantum-mechanical wave–particle duality in which it is neither a particle in the classical sense nor a wave in the classical sense, but a new reality in the quantum world that displays both particle-like and wave-like properties. In non-relativistic quantum mechanics, the probability amplitude for this wave–particle duality is to be determined as solutions of the Schrödinger's equation and in the fully relativistic case as solutions of the Dirac equation. Prior to the development of quantum field theory, the evolution in the concept of particle consisted essentially of two stages: first, Newton's point-mass and then the quantum-mechanical wave-particle duality. This concept of particles would then go through a radical change within the framework of quantum field theory.

One might notice at this point as to why not a single word has been mentioned of the wave equations for electromagnetic fields, which would lead to the equation for photons, the equation that along with the Dirac equation for fundamental fermions completes the founding pillars of quantum field theory. It has not been included up to this point for a very good reason: the wave equation for the electromagnetic field is an equation not of quantum mechanics but of classical physics. The equation for the electromagnetic field predates the advent of both relativity and quantum mechanics. One might go so far as to say that the equation for wave, the electromagnetic wave, has been "waiting" all this while for the equations for particles to "catch up" with it! We will now turn to this classical wave equation for the electromagnetic field in the next chapter.

# 6

# Electromagnetic Field

The classical theory of electromagnetism, as mentioned in Chapter 1, developed along an entirely different path from that of Newton's classical mechanics. From day one, electromagnetism was based on properties of force fields — the electric and magnetic fields that are extended in space. An electric field due to a point charge, for example, is defined over the entire three-dimensional space surrounding the point charge. The works of Coulomb, Gauss, Biot–Savart, Ampère, and Faraday led Maxwell to the great unification of electricity and magnetism into a single theory of an electromagnetic field. Together with Einstein's theory of gravitational field, Maxwell's theory of electromagnetic field is one of the most elegant of classical field theories.

Maxwell's equations are given, in the natural unit system, as

$$\begin{aligned} \nabla \cdot \mathbf{E} &= \rho, \qquad \nabla \times \mathbf{B} - \frac{\partial \mathbf{E}}{\partial t} = \mathbf{J}, \quad \text{(inhomogeneous)} \\ \nabla \cdot \mathbf{B} &= 0, \qquad \nabla \times \mathbf{E} + \frac{\partial \mathbf{B}}{\partial t} = 0. \quad \text{(homogeneous)} \end{aligned}$$

where $\mathbf{E}$ and $\mathbf{B}$ are the electric and magnetic fields and $\rho$ and $\mathbf{J}$ are the electric charge and current densities. The electric and magnetic fields can be expressed in terms of a scalar potential $\phi$ and a vector potential **A** as

$$\mathbf{B} = \nabla \times \mathbf{A}, \qquad \mathbf{E} = -\nabla\phi - \frac{\partial \mathbf{A}}{\partial t},$$

and the two homogeneous Maxwell's equations are satisfied identically.

The electric charge and current densities $\rho$ and $\mathbf{J}$ are components of a single four-vector $J^\mu = (\rho, \mathbf{J})$ and likewise the scalar and vector potentials $\phi$ and $\mathbf{A}$ are components of a four-vector potential $A^\mu = (\phi, \mathbf{A})$. The electric and magnetic fields, $\mathbf{E}$ and $\mathbf{B}$, correspond to components of the antisymmetric electromagnetic field tensor $F^{\mu\nu}$ defined as

$$F^{\mu\nu} \equiv \partial^\mu A^\nu - \partial^\nu A^\mu = \begin{pmatrix} 0 & -E_x & -E_y & -E_z \\ E_x & 0 & -B_z & B_y \\ E_y & B_z & 0 & -B_x \\ E_z & -B_y & B_x & 0 \end{pmatrix}.$$

The electromagnetic field tensor is thus a four-dimensional "curl" of the four-vector potential. In terms of the electromagnetic field tensor, the inhomogeneous Maxwell's equations become

$$\partial_\mu F^{\mu\nu} = J^\nu.$$

We can now draw two very important conclusions about Maxwell's equations. First, the four-potential $A^\mu = (\phi, \mathbf{A})$ is not unique in the sense that the same electromagnetic field tensor $F^{\mu\nu}$ is obtained from the potential

$$A^\mu + \partial^\mu \Lambda = \left(\phi + \frac{\partial \Lambda}{\partial t}, \mathbf{A} - \nabla\Lambda\right),$$

where $\Lambda(x)$ is an arbitrary function and its contribution to $F^{\mu\nu}$ is identically zero (it is the four-dimensional analogue of the curl of gradient being identically zero). This freedom to shift the four-potential $A^\mu = (\phi, \mathbf{A})$ by the four-gradient of an arbitrary function is called *gauge transformation* and it forms the basis for the quantum field theory for the standard model, sometimes also called the theory of gauge fields.

The second conclusion is no less important. The source-free ($J^\mu = 0$) inhomogeneous Maxwell's equations are

$$\partial_\mu F^{\mu\nu} = \partial_\mu \partial^\mu A^\nu - \partial^\nu \partial_\mu A^\mu = 0.$$

Using the freedom of gauge transformation, we can set $\partial_\mu A^\mu = 0$. The choice of the arbitrary function $\Lambda(x)$ to render $\partial_\mu A^\mu$ as always being zero is referred to as the Lorentz gauge. With such an option, Maxwell's equations reduce to

$$\partial_\mu \partial^\mu A^\nu = 0,$$

which, as mentioned in Chapter 1, is exactly the zero-mass case of Klein–Gordon equation.

At the risk of being repetitive, let us emphasize this remarkable point that Maxwell's equations are classical wave equations for the four-potential, and they predate both relativity and quantum mechanics. In this amazing twist, a window has opened up for us to look at the relativistic quantum mechanical wave equations, such as the Klein–Gordon and Dirac equations, in an entirely new light.

# Emulation of Light I: Matter Fields

We are now at the point, after the first six chapters, to look back and compare where the equations of motion for the field and particles stand with respect to each other. As far as electromagnetic fields are concerned, the equations remain intact in its original form, as Maxwell had written down. As discussed in the last chapter, Maxwell's equations for the radiation field, that is, in the source-free region, are of very compact expression. In terms of the four-vector potential and for a particular choice of gauge called the Lorentz gauge, the equations are expressed as

$$\partial_\mu \partial^\mu A^\nu = 0$$

which also coincided with the Klein–Gordon equation for mass-zero case. The equations for particles, on the other hand, evolved through several phases — from Newton to Lagrange and Hamilton and through relativity and quantum mechanics — and ended up in the form of wave equations for non-interacting one-particle within the framework of relativistic quantum mechanics, the Klein–Gordon and Dirac equations being prime examples.

Relativistic quantum mechanical equations as one-particle equations, however, suffer from some serious interpretative problems. For example, the Klein–Gordon equation could not avoid the problem of occurrence of negative probabilities while the Dirac equation suffered from the appearance of negative-energy levels. A new insight was definitely required to proceed to the next phase in the evolution of theories of particles. Such insight would come from the quantization of the electromagnetic field. We will discuss the formalism of the quantization of classical fields in the next chapter. Suffice it to say here that when the radiation field (the electromagnetic field in the source-free region) was quantized, following the recipe for canonical quantization the quantal structure of such quantized radiation field corresponded to photons of Planck and Einstein, the particles of light. This point needs to be repeated: *When a classical field is quantized (in the manner as will be discussed in the next chapter), the quanta of the field are the particles represented by the classical field equation.* This relationship between the classical electromagnetic field and photons, the discreet energy quanta of the radiation field that correspond to particles of light with no mass, provided an entirely new insight into the interpretation of particles. The concept of particles would then go through another fundamental evolution, from that of quantum-mechanical wave–particle duality to that of the quanta of a quantized field.

For particles to be described by relativistic quantum fields, however, there were no corresponding classical fields. We know of only two classical fields in nature, the electromagnetic and gravitational fields. Where and how do we find the classical fields whose quanta correspond to particles satisfying the Klein–Gordon or the Dirac equations? And it is here that we find one of the fundamental conceptual shifts needed to proceed to the next level. *Relativistic quantum-mechanical wave equations such as the Klein–Gordon and Dirac equations are to be reinterpreted as classical field equations at the same level as Maxwell's equation for the classical electromagnetic field!* This is definitely a leap of faith.

Overnight the wave amplitudes for particles (that is, the particle–wave dualities) were turned into corresponding classical fields and the wave equations of relativistic quantum mechanics were turned into corresponding "classical" equations for the classical fields. No equations were modified and all notations remained intact. The wave amplitude $\varphi(x)$ became the classical field $\varphi(x)$ and relativistic quantum-mechanical equations became wave equations for classical fields. This turned out to be one of the most subtle conceptual switches in the history of physics. This was the first instance — it would not be the last — in which matter emulated radiation. This is precisely how we arrived, in the early days of 1930s and 1940s, at the very beginning of quantum field theory of matter — equations for matter simply emulating those for radiation. So, at this point, every wave equation for matter as well as radiation is a classical wave equation for classical fields, some real (Maxwell's equations) and others "imitations" (relativistic quantum mechanical wave equations). We have the truly classical field of the electromagnetic field satisfying

$$\partial_\mu \partial^\mu A^\nu = 0,$$

and the "imitation" classical fields, which are the redressed relativistic quantum-mechanical wave equations, satisfying equations such as those of Klein–Gordon and Dirac, *but now viewed from this point forward as classical field equations*:

$$(\partial^\mu \partial_\mu + m^2)\phi(x) = 0$$

and

$$(i\,\gamma^\mu \partial_\mu - m)\psi_\alpha(x) = 0.$$

Strictly speaking, the classical Klein–Gordon or Dirac field does not exist in the macroscopic scale. No signals are transmitted by these "fields" from one point to another in space in the same way radio signals are carried by the classical radiation field. Reinterpretation of these relativistic quantum-mechanical wave equations as classical field equations is the first preliminary step toward establishing the quantum field theory of matter particles. Once these "imitation" classical fields are quantized in exactly the same manner as the electromagnetic field, the resulting theory of matter particles interacting with photons — quantum electrodynamics — turned out to be the most successful theory for elementary particles to date. In this sense, the redressing of relativistic quantum-mechanical wave equations into "imitation" classical field equations is one of many examples of "the end justifying the means." That is its rationalization.

# 8

# Road Map for Field Quantization

We are now ready to proceed with the quantization of classical fields — the classical electromagnetic, Klein–Gordon and Dirac fields — that were discussed in the last chapter. The quantization of these fields is to be carried out following the rules of canonical quantization, as discussed in Chapter 3. Before imposing canonical quantization onto the classical fields, however, we need to extend the Lagrangian formalism from that of point mechanics to one more suitable for continuous classical field variables.

First and foremost is the question of generalized coordinates . The generalized coordinates $q_i(t)$ with discrete index $i = 1, 2, \ldots, n$ for a system with $n$ degrees of freedom is taken to the limit $n \to \infty$ and in that limit the values of a field at each point of space are to be considered as independent generalized coordinates. Consider a simple mechanical example of a continuous string: the vertical displacement function, say, $\rho(\mathbf{x}, t)$, stands for the amplitude of displacements of the continuous string at position $\mathbf{x}$ and at time $t$ and its values at each position can be taken as independent generalized coordinates. The discrete index $i$ of the generalized coordinates for point mechanics is replaced by the continuous coordinate variable $\mathbf{x}$, and the fields themselves — $A_\mu(x)$ of the electromagnetic field, $\phi(x)$ of the Klein–Gordon field, $\psi_\alpha(x)$ of the Dirac field, and so on — take the place of generalized coordinates.

The canonical formalism for fields requires the canonically conjugate momenta that are to be paired with field variables, and the momenta that canonically conjugate to fields can be defined in terms of the Lagrangian that yields correct equations of motion via Lagrange's equations of motion. In Chapter 2, we took the simplest approach to obtaining Lagrange's equation, starting from Newton's equations of motion. There is another way of obtaining Lagrange's equations that is more formal than the direct approach we took in Chapter 2, and that is to derive Lagrange's equation from what is called Hamilton's principle of least action for particle mechanics. The resulting solution of Hamilton's principle is known mathematically as the Euler equation and Lagrange's equation is a specific example of this more generic Euler equation adopted for particle mechanics. Often, for this reason, Lagrange's equations are also referred to as the Euler–Lagrange equations. We will not get into the details of this formalism here, especially since all that we really need is the expression for Lagrangian that will help define expressions for the momenta canonically conjugate to fields.

For classical fields, it is more convenient to use the Lagrangian densities $\mathcal{L}$ defined as

$$L \equiv \int_{-\infty}^{\infty} d^3x \, \mathcal{L}\left(\phi, \frac{\partial \phi}{\partial x^\mu}\right)$$

and the Euler–Lagrange equations in terms of the Lagrangian densities are given as

$$\frac{\partial}{\partial x^\mu} \frac{\partial \mathcal{L}}{\partial(\partial \phi / \partial x^\mu)} - \frac{\partial \mathcal{L}}{\partial \phi} = 0.$$

Comparing the Euler–Lagrange equations above with the Lagrange's equations given in Chapter 2, we note that the only change is in the leading term where derivatives with respect to time only are replaced by derivatives with respect to all four space-time coordinates $x^\mu$. The momenta conjugate to a field are defined via the Lagrangian density in much the same way as for the case of particle mechanics, thus:

$$\pi(\mathbf{x}, t) \equiv \frac{\partial \mathcal{L}}{\partial(\partial\phi/\partial t)}.$$

We can now state in one paragraph what the quantization of classical fields is all about: (1) start with the classical field equations, Maxwell's, Klein–Gordon, and Dirac equations for radiation field and matter fields, respectively, (2) seek a Lagrangian density for each field that reproduces the field equations via the Euler–Lagrange equations (this is about the only use of Euler–Lagrange equations in this context), (3) with the help of these Lagrangian densities, define momenta canonically conjugate to the fields, and (4) carry out the quantization by imposing commutation relations on these canonically conjugate pairs. After imposing quantization, the fields and their momenta become quantum mechanical operators. *That* sums up in a nutshell what the quantum field theory is all about.

As the fields and their conjugate momenta are both functions of time, as well as of space, they become, upon quantization, operators that are functions of time and this necessarily casts the whole quantum field theory in the Heisenberg picture of quantum theory. As mentioned briefly in Chapter 3, there are two equivalent ways in which the time development of a system can be ascribed to: either operators representing observables or states represented by time-dependent wavefunctions. The former is called the Heisenberg picture and the latter Schrödinger picture. In one-particle quantum mechanics, both non-relativistic and relativistic, it is usually more convenient to adopt the Schrödinger picture and the time development of a system is given by the wavefunctions as functions of time. In quantum field theory wherein the time-dependent fields and their conjugate momenta become operators, the formalism is necessarily in the Heisenberg picture. Let us spell out the bare essence of the relationship between the two pictures, as far as canonical quantization rules are concerned.

In the Schrödinger picture, the wavefunction $\psi(t)$ carries the time development information, that is, if the initial state at an arbitrary time, say $t = 0$, is specified, the Schrödinger's equation determines the state at all future times. The commutation relations for the canonical pairs of operators are, as given before,

$$[q_j, p_k] = i\delta_{jk}$$
$$[q_j, q_k] = [p_j, p_k] = 0.$$

In the Heisenberg picture, the wavefunction is time-independent and is related to that of the Schrödinger picture by

$$\psi_{\mathbf{H}} \equiv \psi_{\mathbf{S}}(0)$$

and the commutators between $q$ and $p$ become, for an arbitrary time $t$,

$$[q_j(t), p_k(t)] = i\delta_{jk}$$
$$[q_j(t), q_k(t)] = [p_j(t), p_k(t)] = 0.$$

In the continuum language of fields and conjugate momenta, they become

$$[\phi(\mathbf{x}, t), \phi(\mathbf{x}', t)] = [\pi(\mathbf{x}, t), \pi(\mathbf{x}', t)] = 0$$
$$[\phi(\mathbf{x}, t), \pi(\mathbf{x}', t)] = i\delta^3(\mathbf{x} - \mathbf{x}')$$

where the Dirac delta function replaces $\delta_{jk}$ and is defined by

$$\int d^3x\, \delta^3(\mathbf{x} - \mathbf{x}') f(\mathbf{x}') = f(\mathbf{x}).$$

In sum, four items — field equations, the Lagrangian densities, momenta conjugate to the fields, and the commutation relations imposed on them — provide the basis of what is called the quantum field theory.

The determination of the Lagrangian density thus plays the very starting point for quantum field theory and Lagrangian densities are so chosen such that the Euler–Lagrange equations reproduce the correct field equations for a given field. The choice, however, is not unique since the Euler–Lagrange equations involve only the derivatives of the Lagrangian densities. A Lagrangian density is chosen to be the simplest choice possible that meets the requirement of reproducing the field equations when substituted into the Euler–Lagrange equations. The Lagrangian densities are:

$$\mathcal{L} = \frac{1}{2}(\partial_\mu \phi \, \partial^\mu \phi - m^2\phi^2) \quad \text{for the Klein–Gordon field,}$$

$$\mathcal{L} = -\frac{1}{4}F_{\mu\nu}F^{\mu\nu} \quad \text{for the electromagnetic field, and}$$

$$\mathcal{L} = \bar{\psi}(i\gamma^\mu \partial_\mu - m)\psi \quad \text{for the Dirac-field}$$

where $\psi$ is the four-component Dirac field (column) and $\bar{\psi}$ is defined as $\bar{\psi} = \psi^* \gamma^0$, an adjoint (row) multiplied by $\gamma^0$, referred to as the Dirac adjoint, which is simply a matter of notational convenience that became a standard notation.

The canonical quantization procedure in terms of the commutators, as shown above, is rooted in the Poisson bracket formalism of Hamilton's formulation of mechanics, as discussed in Chapter 3. It leads to a successful theory of quantized fields for the Klein–Gordon and electromagnetic fields, that is, those that represent particles of spin zero and one, in fact, of all integer values of spin. The particles of half-integer spins, half, and one and half, and so on, must satisfy the Pauli exclusion principle and the fields that represent these particles, the Dirac field in particular, must be quantized not by commutators

$$[A, B] \equiv AB - BA$$

but by anticommutators

$$\{A, B\} \equiv AB + BA.$$

The choice of commutators versus anticommutators depending on whether the spin has integer or half-integer value can be compactly expressed as

$$AB - (-1)^{2s}BA$$

where $s$ stands for either integer or half-integer. The anticommutators have no classical counterparts, that is, there are no such things as Poisson antibrackets, but nevertheless the quantization by anticommutators is one of the fundamental requirements for quantizing fields that correspond to particles of half-integer spins. In the case of Dirac fields, the origin of the use of anticommutators can be traced back to $\gamma^\mu$ matrices that are required, by definition, to satisfy anticommutation relations among them.

Another item to be mentioned here is concerned with the wide use of the term "second quantization." When referring to the quantization of matter fields, the term accurately describes the situation. The equations for matter fields — Klein–Gordon and Dirac fields — are actually relativistic quantum mechanical wave equations. That is the "first" quantization. The wave equations are then viewed as "classical" field equations, in an emulation of the classical electromagnetic field, and then quantized again. That is the "second" quantization. As far as the classical wave equations for electromagnetic fields are concerned, however, this is not accurate. For the electromagnetic field, the wave equation is a classical wave equation and its quantization is its "first" quantization. The term "second quantization," however, picked up a life of its own and became synonymous with quantum field theory and is widely used interchangeably with the latter. Strictly speaking, such interchangeable use of the two terms is not entirely accurate, especially where quantization of the classical electromagnetic field is concerned.

# 9

# Particles and Fields III: Particles as Quanta of Fields

Quantum field theory presents the third and the latest stage in the evolution of the concept of particles. This concept has evolved from that of a localized point mass in classical physics to that of wave–particle duality in quantum mechanics and, as shown in this chapter, to that of a quantum of quantized field. As classical fields are quantized, following the road map outlined in the last chapter, we will see that the concept of particles has become secondary to that of quantized fields. The quantal structure of fields, or more precisely the quantal structure of energy and momentum of fields, defines particles as discrete units of the field carrying the energy and momentum characteristics of each particle. In this sense, fields play the primary physical role and particles only the secondary role as units of discrete energy of a given field.

When the electromagnetic four-potential $A_\mu(x)$ is quantized, according to standard procedure outlined in the previous chapter, it becomes a field operator, much the same way that $q$'s and $p$'s, the coordinates and momenta, turn into operators in quantum mechanics. The field operator $A_\mu(x)$ consists of two parts — this is common property for all fields, whether Klein–Gordon or Dirac — called the positive frequency and negative frequency parts. The positive frequency part corresponds to raising the energy of an electromagnetic system by one unit of quantum and this corresponds to the creation operator of a photon. Likewise, the negative energy frequency part corresponds to lowering the energy of an electromagnetic system by one unit of quantum and this is the annihilation operator of a photon. The successful incorporation of photons, the zero-mass particles of light, into the fold of quantized electromagnetic four-potential was in fact the catalyst, as discussed in previous chapters, for the reinterpretation of relativistic wave equations of particles as "imitation" matter fields which started the whole ball rolling toward today's quantum field theory of particles.

The description of quantization of fields — electromagnetic, Klein–Gordon, Dirac and others — is the first-order of business for any graduate-level textbooks of quantum field theory and is usually featured in a substantial part of such textbooks. Quantization of the Dirac field alone, for example, usually takes two to three long chapters to discuss all the relevant details. We will refer to any one of the standard textbooks for full details of field quantization[1] and strive only to bring out its essential aspects in this chapter. In order to illustrate the emergence of the creation and annihilation operators of the field operators, it is very helpful first to briefly review the operator techniques involved in the case of simple harmonic oscillator problem of non-relativistic quantum mechanics.

Consider a one-dimensional simple harmonic oscillator. Denote the energy of the system by $H$ (for Hamiltonian, which is equal to the total energy) as

$$H = \frac{p^2}{2m} + \frac{1}{2}m\omega^2 x^2$$

where $m$ is the mass of the oscillating particle and $\omega$ is the classical frequency of oscillation. The expression above for energy can also be

[1]For example, such classics as *An Introduction to Relativistic Quantum Field Theory* by S. Schweber (1961), *Relativistic Quantum Fields* by J. Bjorken and S. Drell (1965), and *Introduction to Quantum Field Theory* by P. Roman (1969).

denoted as

$$H = \left(a^*a + \frac{1}{2}\right)\omega = \left(aa^* - \frac{1}{2}\right)\omega$$

where (recall that $\hbar = 1$ in the natural unit system)

$$a = \frac{1}{\sqrt{2m\omega}}(ip + m\omega x)$$
$$a^* = \frac{1}{\sqrt{2m\omega}}(-ip + m\omega x).$$

The basic commutation relation between $x$ and $p$

$$[x, p] = i$$

leads to the following commutation relations,

$$[a, a] = [a^*, a^*] = 0$$

and

$$[a, a^*] = 1.$$

As is well known from elementary non-relativistic quantum mechanics, the lowest energy (the ground state) of the system is equal to $\frac{1}{2}\omega$ and the operators $a^*$ and $a$, respectively, increase or decrease energy by the quantized amount equal to $\omega$. The quantized energy in units of $\omega$ represents the quantum of the system and the operators $a^*$ and $a$, respectively, raise and lower the energy by the oscillator. For this reason $a^*$ and $a$, are respectively called the raising and lowering operators, for a simple harmonic oscillator.

When we quantize a field, an exactly analogous situation occurs: the field operator consists of "raising" and "lowering" operators that increase or decrease the energy of the system described by the field and it is the "quantum" of that energy that corresponds to the particle described by the field. The "raising" and "lowering" operators of the field are called creation and annihilation operators and the "quantum" of energy corresponds to a particle described by the field. Although it was the quantization of electromagnetic field that preceded that of matter fields, we will illustrate this procedure for the simplest case: Klein–Gordon field which is a scalar field (spin zero) and real (complex fields describe charged spin zero particles). The quantization of electromagnetic and Dirac fields has the added complications of having to deal with spin indices for photons (polarizations of photons) and half-integer spin particles such as electrons.

The case of Klein–Gordon field is specified by, as discussed in Chapter 8:

| | |
|---|---|
| Field: | $\phi(x)$ |
| Field equation: | $(\partial^\mu \partial_\mu + m^2)\phi(x) = 0$ |
| Lagrangian density: | $\mathcal{L} = \frac{1}{2}(\partial_\mu \phi \partial^\mu \phi - m^2 \phi^2)$ |
| Momenta field: | $\pi(\mathbf{x}, t) \equiv \frac{\partial \mathcal{L}}{\partial(\partial \phi / \partial t)} = \frac{\partial \phi}{\partial t}.$ |

The field equation, that is, the Klein–Gordon equation, allows plane-wave solutions for the field $\phi(x)$ and it can be written as

$$\phi(x) = \frac{1}{(2\pi)^{3/2}} \int b(k) e^{ikx}\, dk$$

where $kx = k^0 x^0 - \mathbf{kr}$, $dk = dk^0 d\mathbf{k}$ and $b(k)$ is the Fourier transform that specifies particular weight distribution of plane-waves with different $k$'s. As a solution of the field equation, there is a restriction on the transform $b(k)$, however. Substituting the plane-wave solution into the field equation shows that $b(k)$ has the form

$$b(k) = \delta(k^2 - m^2) c(k)$$

in which $c(k)$ is arbitrary. The delta function simply states that as the solution of Klein–Gordon equation, the plane-wave solution must obey Einstein's energy–momentum relation, $k^2 - m^2 = 0$.

The Einstein's energy–momentum relation ($k^2 - m^2 = 0$) also places a constraint on $dk$, and as we shall see, this constraint is one of the basic ingredients of quantization of all fields. The integral over $dk$ is not all over the $k^0 - \mathbf{k}$ four-dimensional space, but rather only over $d\mathbf{k}$ with $k^0$ restricted by the relation, (for notational convenience we switched from $(k^0)^2$ to $k_0^2$)

$$k_0^2 - \mathbf{k}^2 - m^2 = 0.$$

Introducing a new notation

$$\omega_k \equiv +\sqrt{\mathbf{k}^2 + m^2} \quad \text{with only the + sign,}$$

either $k_0 = +\omega_k$ or $k_0 = -\omega_k$. Integrating out $k_0$, the plane-wave solutions decompose into "positive frequency" and "negative frequency" parts.[2] This decomposition, which is basic to all relativistic fields, matter fields as well as the electromagnetic field, has nothing to do with field quantization and is rooted in the quadratic nature of Einstein's energy–momentum formula. The plane-wave solutions can be written in the form:

$$\phi(x) = \int d^3\mathbf{k}(a(\mathbf{k}) f_k(x) + a^*(\mathbf{k}) f_k^*(x))$$

where

$$f_k(x) = \frac{1}{\sqrt{(2\pi)^3 2\omega_k}} \mathrm{e}^{-ikx} \quad \text{and} \quad f_k^*(x) = \frac{1}{\sqrt{(2\pi)^3 2\omega_k}} \mathrm{e}^{+ikx}.$$

The integral is over $d^3\mathbf{k}$ only and $a(\mathbf{k})$ and $a^*(\mathbf{k})$ are the respective Fourier transforms for "positive frequency" and "negative frequency" parts. Two remarks about the standard practice of notations are called for here: The star superscript (*) stands for complex conjugate in classical fields, but when they are quantized and become non-commuting operators, the notation will stand for Hermitian adjoint. After decomposition into "positive frequency" and "negative frequency" parts, the notation $k_0$, as in $\mathrm{e}^{-ikx}$, stands as a shorthand for $+\omega_k$, that is, after $k_0$ is integrated out, notation $k_0 = +\omega_k$. For

[2]See Appendix 4 for more details.

brevity, we often write

$$\phi(x) = \phi^{(+)}(x) + \phi^{(-)}(x)$$

with

$$\phi^{(+)} = \int d^3\mathbf{k} a(\mathbf{k}) f_k(x) \quad \text{and} \quad \phi^{(-)}(x) = \int d^3\mathbf{k} a^*(\mathbf{k}) f_k^*(x).$$

We now quantize the field by imposing the canonical quantization rule, mentioned in Chapter 8, namely:

$$[\phi(\mathbf{x},t), \phi(\mathbf{x}',t)] = [\pi(\mathbf{x},t), \pi(\mathbf{x}',t)] = 0$$
$$[\phi(\mathbf{x},t), \pi(\mathbf{x}',t)] = i\delta^3(\mathbf{x} - \mathbf{x}')$$

where $\pi(\mathbf{x},t) \equiv \partial\mathcal{L}/\partial(\partial\phi/\partial t) = \partial\phi/\partial t$. These commutation rules become commutation relations among $a(\mathbf{k})$'s and $a^*(\mathbf{k})$'s, thus:

$$[a(\mathbf{k}), a(\mathbf{k}')] = [a^*(\mathbf{k}), a^*(\mathbf{k}')] = 0$$
$$[a(\mathbf{k}), a^*(\mathbf{k}')] = \delta^3(\mathbf{k} - \mathbf{k}').$$

These commutation relations are essentially identical to those for raising and lowering operators of the simple harmonic oscillator. The quantal structure of the quantized field, and resulting new interpretation of particles, is then exactly analogous to the case of raising and lowering operators for a simple harmonic oscillator.

We define the vacuum state (no-particle state), $\Psi_0$, to be the state with zero energy, zero momentum, zero electric charge, and so on. When we operate on this vacuum state with operator $a^*(\mathbf{k})$, the resulting state

$$\Psi_1 \equiv a^*(\mathbf{k})\Psi_0 \text{ (one-particle state)}$$

corresponds to that with one "quantum" of the field that has a momentum $\mathbf{k}$ and energy $\omega_k \equiv +\sqrt{\mathbf{k}^2 + m^2}$. With $E = \omega$ and $\mathbf{p} = \mathbf{k}$ ($\hbar = 1$), this quantum is none other than a relativistic particle of mass $m$ defined by

$$E^2 - \mathbf{p}^2 = m^2.$$

A spin zero particle of mass $m$ thus corresponds to the quantum of Klein–Gordon field and is created by the operator $a^*(\mathbf{k})$. For this reason, the operator $a^*(\mathbf{k})$ is called the creation operator. The operator

$a(\mathbf{k})$ does just the opposite,

$$a(\mathbf{k})\Psi_1 = \Psi_0,$$

and the operator $a(\mathbf{k})$ is called the annihilation operator. Repeated application $a^*(\mathbf{k})$'s leads to two, three, ..., $n$-particle state; likewise repeated application of $a(\mathbf{k})$'s reduces the number of particles from a given state. The quantized Klein–Gordon field operator hence contains two parts, one that creates a particle and the other that annihilates a particle: a field operator acting on the $n$-particle state gives both $(n+1)$- and $(n-1)$-particle states. A relativistic particle of mass $m$ now corresponds to the quantum of the quantized field. This is the third and, so far the final, evolution in our concept of a particle.

The quantization of an electromagnetic field is virtually identical to that discussed above for the Klein–Gordon field, except that due to the polarization degrees of freedom (the spin of photons), the field $A_\mu(x)$ requires a little more care. The polarization of the electromagnetic field has only two degrees of freedom, the right-handed and left-handed circular polarizations, but the field $A_\mu(x)$ has four indices ($\mu = 0, 1, 2, 3$). This is usually taken care of by making a judicial choice allowed by gauge transformation: we choose such a gauge in which $A_0 = 0$ and $\nabla \cdot \mathbf{A} = 0$. This choice, called the radiation gauge, renders $A_\mu(x)$ to have only two independent degrees of freedom. Besides the added complications involved in the description of polarizations, the remaining procedures in the quantization of electromagnetic field are identical to that of the Klein–Gordon field and, after quantization, the electromagnetic field operator also decomposes into creation and annihilation parts:

$$\begin{aligned} A_\mu(x) &= \text{annihilation operator for a single photon} \\ &\quad + \text{creation operator for a single photon} \\ &= A_\mu^{(+)}(x) + A_\mu^{(-)}(x). \end{aligned}$$

As the quantum of electromagnetic field, the photon is a particle that has zero mass and carries energy and momentum given by $E = |\mathbf{p}| = \omega(\hbar = c = 1)$.

The quantization of Dirac field is more involved on several accounts: first, the Dirac field $\psi$ is a four-component object, as discussed in Chapter 5, the Lagrangian density involves not only the field $\psi$ but also its Dirac adjoint $\bar{\psi} = \psi^* \gamma^0$, and the canonical quantization must be carried out in terms of anticommutators rather than the usual commutators, as discussed in Chapter 8. When all is said and done, the Dirac field operators decompose as follows (say, for the electron):

$$\psi(x) = \psi^{(+)}(x) + \psi^{(-)}(x)$$

$$\psi^{(+)}(x) \quad \text{annihilates an electron}$$

$$\psi^{(-)}(x) \quad \text{creates a positron (anti–electron)}$$

and

$$\bar{\psi}(x) = \bar{\psi}^{(+)}(x) + \bar{\psi}^{(-)}(x)$$

$$\bar{\psi}^{(+)}(x) \quad \text{annihilates a positron}$$

$$\bar{\psi}^{(-)}(x) \quad \text{creates an electron.}$$

To sum up, when we quantize a field, it turns into a field operator that consists of creation and annihilation operators of the quantum of that field. In the case of an electromagnetic field, the classical field of the four-vector potential turns into creation and annihilation operators for the quantum of that field, the photon. In the case of matter fields, we first reinterpret the one-particle relativistic quantum mechanical wave equations as equations for classical matter fields and then carry out the quantization. Matter particles, be they spin zero scalar particles or spin half particles such as electrons, positrons, protons and neutrons, emerge as the quanta of quantized matter fields, whether they are Klein–Gordon or Dirac fields. Essentially, this is what quantum field theory of particles is all about.

# 10

# Emulation of Light II: Interactions

The quantization of fields and the emergence of particles as quanta of quantized fields discussed in Chapter 9 represent the very essence of quantum field theory. The fields mentioned so far — Klein–Gordon, electromagnetic as well as Dirac fields — are, however, only for the non-interacting cases, that is, for free fields devoid of any interactions, the forces. The theory of free fields by itself is devoid of any physical content: there is no such thing in the real world as a free, non-interacting electron that exerts no force on an adjacent electron. The theory of free fields provides the foundation upon which one can build the framework for introducing real physics, namely, the interaction among particles.

We must now find ways to introduce interactions into the procedure of canonical quantization based on the Lagrangian and Hamiltonian formalism. The question then is what is the clue and prescription by which we can introduce interactions into the Lagrangian densities. There are very few clues. In fact, there is only one known prescription to introduce electromagnetic interactions and it comes from the Hamiltonian formalism of classical physics, as discussed in Chapter 2. Comparing the classical Hamiltonian (total energy) for a free particle with that of the particle interacting with the electromagnetic field, the recipe for introducing the electromagnetic interaction is the substitution rule (sometimes referred to as the "minimal" substitution rule)

$$p_\mu \Rightarrow p_\mu - eA_\mu.$$

Replacing $p_\mu$ by its quantum-mechanical operator $i\partial_\mu$, we have as the only known prescription for introducing electromagnetic interaction:

$$i\partial_\mu \rightarrow i\partial_\mu - eA_\mu.$$

We switched the notation for the charge from $q$ to $e$. This substitution is to be made only in the Lagrangian density of free matter fields representing charged particles, but not to every differential operator that appears in a given Lagrangian density, not, for example, to differential operators in the Lagrangian density for a free electromagnetic field.

In the last chapter, we used the simple scalar Klein–Gordon field to illustrate the process of field quantization and the resulting emergence of particles as quanta of the field. To illustrate the introduction of interaction by substitution rule, we switch from Klein–Gordon to the Dirac field. All particles of matter — electrons, protons, neutrons that make up atoms, that is, all quarks and leptons (more on these later) — are spin half particles satisfying the Dirac field equations and the description of electromagnetic interactions of these particles, say, electron, requires the substitution rule to be applied to the Lagrangian density for the Dirac field.

The Lagrangian density for charged particles, say, electrons, interacting with the electromagnetic field is then given by applying the substitution rule to the Lagrangian density for the free Dirac field, and combining with the Lagrangian density for the electromagnetic field, we have

$$\begin{aligned}\mathcal{L} &= \bar{\psi}(\gamma^\mu(i\partial_\mu - eA_\mu) - m)\psi - \frac{1}{4}F_{\mu\nu}F^{\mu\nu} \\ &= \bar{\psi}(i\gamma^\mu\partial_\mu - m)\psi - \frac{1}{4}F_{\mu\nu}F^{\mu\nu} - e\bar{\psi}\gamma^\mu\psi A_\mu.\end{aligned}$$

For brevity, we omitted the functional arguments, $(x)$, from all fields in the above expression, i.e. $\mathcal{L} = \mathcal{L}(x)$, $\psi = \psi(x), A_\mu = A_\mu(x)$, etc. The new Lagrangian describes the local interaction of electron and photon fields at the same space–time point $x$. Substituting this interaction Lagrangian into the Euler–Lagrange equation, we obtain the field equations for interacting fields, which as expected, are different from the equations for free Dirac and electromagnetic fields:

$$\begin{aligned}(i\gamma^\mu \partial_\mu - m)\psi(x) &= e\gamma^\mu A_\mu \psi(x)\\ \partial_\nu F^{\mu\nu} &= e\bar{\psi}(x)\gamma^\mu \psi(x).\end{aligned}$$

We need to make several important observations here about this new interaction Lagrangian. First, the field equations for interacting fields are highly nonlinear and they are also coupled; to solve one, the other must be solved. The Dirac and electromagnetic fields, $\psi(x)$ and $A_\mu(x)$, that appear in the interaction Lagrangian, although they have the same notation, are *not* the same as the free Dirac and electromagnetic fields. Secondly, in order to proceed with the quantization of interacting fields, as illustrated in the case of free fields in the last chapter, the first thing we need are the solutions to the coupled field equations given above. We could then presumably proceed to decompose the solutions for interacting fields and perhaps even define "interacting particle creation and annihilation operators." Once we have the exact and analytical solutions for fields satisfying the coupled equations, we may have the emergence of real, physical particles as quanta of interacting fields. Quantum field theory for interacting particles would have been completely solved, and we could have moved on beyond it. Well, not exactly. Not exactly, because no one can solve the highly nonlinear coupled equations for interacting fields that result from the interacting Lagrangian density obtained by the substitution rule. Exact and analytical solutions for interacting fields have never been obtained; we ended up with the Lagrangian that we could not solve!

Just to illustrate one key point of departure from the quantization of free fields, consider the requirement that each component of the free Dirac field must also satisfy, over and beyond the Dirac equation, the Klein–Gordon equation. The requirement that $k_0^2 - \mathbf{k}^2 - m^2 = 0$, that is, $k_0 = +\omega_k$ or $k_0 = -\omega_k$ with $\omega_k \equiv +\sqrt{\mathbf{k}^2 + m^2}$, is what allowed the decomposition of the free field into creation and annihilation operators, that, in turn, led to particle interpretation. In the case of interacting fields, this is no longer possible.

At this point, the quantum field theory of interacting particles proceeded towards the only other alternative left: when so justified, treat the interaction part of the Lagrangian as a small perturbation to the free part of the Lagrangian. We write

$$\mathcal{L}(x) = \mathcal{L}_{\text{free}}(x) + \mathcal{L}_{\text{int}}(x)$$

where

$$\begin{aligned}\mathcal{L}_{\text{free}}(x) &= \bar{\psi}(x)(i\gamma^\mu \partial_\mu - m)\psi(x) - \frac{1}{4}F^{\mu\nu}(x)F_{\mu\nu}(x)\\ \mathcal{L}_{\text{int}}(x) &= -e\bar{\psi}(x)\gamma^\mu \psi(x)A_\mu(x).\end{aligned}$$

The perturbative approach with the Lagrangian above is the basis for quantum field theory of charged particles interacting with the electromagnetic field, to wit, the quantum electrodynamics, QED. To this date, QED, with some further fine-tuning (more on this in the next chapter), remains the most successful — and so far the only truly successful — theory of interacting particles ever devised. The perturbative approach of QED is well justified by the smallness of the charge, $e$, renamed the coupling constant (the fine-structure constant defined as $\alpha = e^2/4\pi$ is approximately equal to $1/137$), which ensures that successive higher orders of approximation would be smaller and smaller. In the zeroth-order, then, the total Lagrangian is equal to free Lagrangian and by the same token, in the zeroth-order, the interacting fields are equal to free fields, and successive orders in the perturbation expansion in terms of the interaction Lagrangian add "corrections" to this zeroth-order approximation, generically called the radiative corrections.

The interaction Lagrangian,

$$\mathcal{L}_{\text{int}}(x) = -e\bar{\psi}(x)\gamma^\mu \psi(x)A_\mu(x),$$

is thus the centerpiece of QED. It is a compact expression that contains, interpreted in terms of the free fields, eight different terms involving various creation and annihilation operators, for each index $\mu$, for a total of 32 terms. For each $\mu = 0, 1, 2, 3$, the expression

$$j^{\mu} = \bar{\psi}(x)\gamma^{\mu}\psi(x),$$

which is a four-element row matrix times a $4 \times 4$ matrix times a four-element column matrix, expands to [creation of electron + annihilation of positron] multiplied by [creation of positron + annihilation of electron]. This $j^{\mu} = \bar{\psi}(x)\gamma^{\mu}\psi(x)$ is then multiplied by [creation of photon + annihilation of photon], three field operators being coupled at the same space–time point $x$.

The success of QED, albeit by the perturbative approach, has catapulted to the above form of interaction Lagrangian to much greater significance and is more fundamental than originally perceived; it became the mantra for all other interactions among elementary particles, namely, the weak and strong nuclear forces. The weak and strong nuclear forces, as well as the electromagnetic force, are to be written in the form

$$g\bar{\Psi}(x)\gamma^{\mu}\Psi(x)B_{\mu}(x)$$

where $g$ is the generic notation for coupling constants, be it electromagnetic, weak nuclear or strong nuclear force, $B_{\mu}(x)$ is the generic notation for the force field of each force, and the Dirac field operators for all spin half matter fields. This expression forms the basis of our understanding of all three interactions at a local point and hence, by extension, the microscopic nature of these forces — three field operators — Dirac field, Dirac adjoint field, and the force field operators — all come together at a space–time point $x$.

For nonelectromagnetic interactions, weak and strong nuclear forces, the adoption of the interaction Lagrangian modeled after the electromagnetic interaction Lagrangian is basically a matter of faith and can be justified only by the success of just extension. We assume the interactions to be derivable from Lagrangian density (this assumption gets some degree of justification when viewed in terms of the so-called "gauge" fields, as will be discussed in a later chapter) and to be just as local as the electromagnetic interaction. Lacking any theoretical basis, such as the substitution rule in the case of electromagnetic interaction, the casting of non-electromagnetic interactions in the form of the interaction Lagrangian density given above corresponds to a grand emulation of the electromagnetic force, to wit, an emulation of light indeed.

# 11

# Triumph and Wane

The success of quantum electrodynamics in agreeing with and predicting some of the most exact measurements is nothing less than spectacular. The quantitative agreements between calculations of QED and experimental data for such atomic phenomena as the Lamb shift, the hyperfine structure of hydrogen, and the line shape of emitted radiation in atomic transitions are truly impeccable and has helped to establish QED as the most successful theory of interacting particles. As stated previously, this is what made QED the shining example to emulate for other interactions.

To proceed from the interaction Lagrangian density

$$\mathcal{L} = \bar{\psi}(i\gamma^\mu \partial_\mu - m)\psi - \frac{1}{4}F_{\mu\nu}F^{\mu\nu} - e\bar{\psi}\gamma^\mu \psi A_\mu$$

to the results of calculations that are in remarkable agreement with observation, however, the theory had to be negotiated through some tortuous paths — calculations that yield infinities, the need to redefine some parameters that appear in the Lagrangian density, and proof that all meaningless infinities that occur can be successfully absorbed in the redefinition program. They are respectively called the ultraviolet divergences, mass and charge renormalizations, and renormalizability of QED.

With the Lagrangian density, and the resulting highly coupled field equations, that could not be solved exactly, there was only one recourse left and that was to seek approximate solutions in which the interaction term was treated as a perturbation to the free-field Lagrangian. The smallness of the coupling constant $e$ would seem to ensure that such perturbation approach is amply justified. But when calculations were carried out order by order in the perturbation expansion in terms of the interaction Lagrangian, the results were disastrous; calculations led to results that were infinite!

The origin of infinities is believed to be an inherent property of the canonical formalism of field theory; within the Lagrangian framework, the values of a field at every space–time point $x$ is considered as generalized coordinates and clearly there are infinite number of generalized coordinates. For example, consider a system consisting of an infinite number of non-relativistic quantum mechanical harmonic oscillators. The ground state energy of each oscillator is $1/2\ \omega$, but the total energy of the system is, of course, infinite. As a system of infinite number of generalized coordinates, appearance of infinities in calculations is actually not surprising. The appearance of infinities is called the problem of ultraviolet divergences. The way to get around this near fatal situation is in what is called the mass and charge renormalizations.

As discussed in the last chapter, the Lagrangian density breaks up into two parts:

$$\mathcal{L}(x) = \mathcal{L}_{\text{free}}(x) + \mathcal{L}_{\text{int}}(x)$$

with

$$\begin{aligned}\mathcal{L}_{\text{free}}(x) &= \bar{\psi}(x)(i\gamma^{\mu}\partial_{\mu} - m)\psi(x) - \frac{1}{4}F^{\mu\nu}(x)F_{\mu\nu}(x)\\ \mathcal{L}_{\text{int}}(x) &= -e\bar{\psi}(x)\gamma^{\mu}\psi(x)A_{\mu}(x).\end{aligned}$$

In the perturbation approach, we imagine the interaction Lagrangian to be switched off, in the zeroth-order approximation, and are then left with the well-established free field theory. Of course, in reality this cannot be true, no more than for us to claim that we have an electron without electromagnetic interaction! Now, there are two basic parameters that enter into the total Lagrangian, the mass $m$ and the charge $e$. Within the perturbation approach, they represent the mass and charge of a totally hypothetical electron that has no electromagnetic interaction. The mass and charge parameters in the Lagrangian cannot be the actual, physically measured mass and charge of a real, physical electron. They must be recalibrated so as to correspond to the measured values of mass and charge. This need to recalibrate the two fundamental parameters that appear in the Lagrangian density is called renormalization, mass and charge renormalizations.

The requirements of mass and charge renormalizations, on the one hand, and the inescapable appearance of infinities in perturbative calculations, on the other hand, are actually quite separate issues; they trace their origins to different sources. In practice, however, the two become inseparably intertwined in that we utilize the procedures to renormalize mass and charge to absorb, and thus get rid of, the unwanted appearance of infinities in calculations. We refer to the mass parameter that appear in the Lagrangian as the *bare* mass, of an electron, and change its notation from $m$ to $m_0$and define the physically observed mass, of an electron, as $m$. The physical mass is then related to the bare mass by

$$m = m_0 - \delta m.$$

Both $m_0$ and $\delta m$ are unmeasurable and unphysical quantities. The physically measured mass of an electron, 0.5 MeV, corresponds to the physical mass $m$ defined as the difference between the bare mass and $\delta m$, sometimes called the mass counter term. In situations where no infinities appear, the mass counter term should be, in principle, calculable from the interaction Lagrangian. It can then be shown in the perturbation calculations that certain types of infinities that occur can all be lumped into the mass counter term. With the bare mass also taken to be of infinite value, the two infinities — the infinities coming out of the perturbation calculations and the infinity of the bare mass — cancel each other out leaving us with a finite value for the actual, physical mass of an electron. The difference between two different infinities can certainly be finite. This process, quite fancy indeed, is called mass renormalization.

The procedure for charge renormalization is a bit more involved than that for mass renormalization, but the methodology is the same. The physical charge is the finite quantity that results from the cancellation of two infinities — between the infinite bare charge and certain other types of infinities that appear in the calculations, that is, types other than those absorbed in mass renormalization.

The crucial test is to show that all types of infinites that occur in the perturbation calculations can be absorbed by the recalibration procedure of physical parameters, that is, the mass and charge renormalizations. Then, and only then, solutions obtained by the perturbation expansion can be accepted. This acid requirement is called the renormalizability of a theory. The two critical requirements for a quantum theory of interacting fields are thus:

(i) Perturbation expansion in terms of the interaction Lagrangian must be justified in terms of the smallness of the coupling constant.
(ii) Such expansion is proven to be renormalizable.

QED passes these two requirements with flying colors. The question now is what about the non-electromagnetic interactions.

As discussed in the last chapter, the interaction Lagrangian for the weak and strong nuclear forces was obtained simply by emulating the format for the electromagnetic interaction. Lacking any specific guide such as the substitution rule, which is deeply rooted in the Lagrangian and Hamiltonian formalism of classical physics, all we could do for these non-electromagnetic forces was to adopt the interaction form given by

$$g\bar{\Psi}(x)\gamma^{\mu}\Psi(x)B_{\mu}(x)$$

where $g$ is the coupling constant signifying the strength of force, $\Psi(x)$ is the relevant Dirac field — proton, neutron, electron and other Dirac fields — and $B_{\mu}(x)$ is the spin one force field. We immediately run into a brick wall when it comes to the strong nuclear interaction: the coupling constant is too large for perturbation expansion in terms of the interaction Lagrangian to be considered. In the same scale as the fine structure constant $\alpha = e^2/4\pi$ of the electromagnetic interaction being equal to 1/137, the coupling constant for the strong nuclear interaction is approximately equal to 1. The question of the perturbation expansion in terms of the coupling constant simply goes out the window for a strong nuclear force. For a weak nuclear interaction, the problem is opposite. The coupling constant for the weak nuclear force is small enough, much smaller in fact than the fine structure constant, and this in itself would ensure the validity of perturbation expansion. Rather, the problem was renormalizability. The number of infinities that occur in the perturbation calculations far exceeded the number of parameters that could absorb them by renormalization. The theory as applied to the weak interaction was simply non-renormalizable. Spectacular triumph was noted in the case of the electromagnetic interaction on the one hand, and complete failures in the case of weak and strong nuclear interactions on the other hand. In the early 1950s, this was the situation.

Quantum field theory cast in the framework of canonical quantization — often called the Lagrangian field theory — came to its mixed ending, unassailable success of QED followed by non-expandability in the case of strong nuclear force and by non-renomalizability in the case of weak nuclear force. And thus ended what might be called the first phase of quantum field theory, the era of success of the Lagrangian field theory in the domain of electromagnetic interaction with the attempt to emulate the success of QED for the case of weak and strong nuclear forces ending up in total failure.

Starting from the 1950s, interest in the Lagrangian field theory thus began to wane and the need to make a fresh start became paramount. This was the beginning of what may be called the second phase of quantum field theory. Discarding the doctrine of the canonical quantization within the Lagrangian and Hamiltonian framework, new approaches were adopted to construct an entirely new framework: building on basic sets of axioms and symmetry requirements, constructing scattering matrices for incorporating interactions that could relate to the observed results. There have been many branches of approach in this second phase, often referred to as the axiomatic quantum field theory, and they occupied a good part of two decades,

1950s and 1960s. But in the end, the axiomatic quantum field theory could not bring us any closer to analytic solutions for interacting fields. By the end of 1960s, the hope for formulating a successful quantum field theory for non-electromagnetic forces began to dim.

Beginning with the 1970s, however, a new life was injected for the Lagrangain field theory — a new perspective on how to introduce the electromagnetic interaction and a new rationale for emulating it for non-electromagnetic interactions. It is called the "local gauge field theory." Coupled with the newly-gained knowledge of what we consider to be the ultimate building blocks of matter, this new local gauge field theory would come to define what we now call the "standard model" of elementary particles. The advent of local gauge Lagrangian field theory is the latest in the development of quantum field theory and corresponds to what may be called its third phase — canonical Lagrangian, axiomatic, and now the local gauge Lagrangian field theory.

# 12

# Emulation of Light III: Gauge Field

As mentioned in the last chapter, the heyday of quantum electrodynamics was over by the early 1950s and in the next two decades, the 1950s and 1960s, the canonical Lagrangian field theory was rarely spoken of. The 50s and 60s were primarily occupied by the search for patterns of symmetries in the world of elementary particles — such discoveries as strangeness, charm, unitary symmetry, the eightfold way, the introduction of quarks, and many others — and the pursuit of quantum field theory was carried out by those investigating the formal framework of the theory, generally called the axiomatic field theory, starting from scratch seeking new ways to deal with weak and strong nuclear forces. During this period that may be called the second phase of quantum field theory, the Lagrangian field theory was almost completely sidelined and the emphasis was on the formal and analytic properties of scattering matrix, the so-called S-matrix theories and the axiomatic approaches to field theory. These new axiomatic approaches, however, did not bring solutions to quantum field theories any closer than the Lagrangian field theories. Entering the 1970s, there was a powerful revival of the Lagrangian field theory that continues to this day. This is what is called the (Lagrangian) gauge field theory, and it starts — yes, once again — from electrodynamics! The gauge field theory represents the third and current phase in the development of quantum field theory.

The Lagrangian density for QED, obtained by the substitution rule,

$$\mathcal{L} = \bar{\psi}(i\gamma^{\mu}\partial_{\mu} - m)\psi - \frac{1}{4}F_{\mu\nu}F^{\mu\nu} - e\bar{\psi}\gamma^{\mu}\psi A_{\mu}$$

is clearly invariant under a phase change of the field $\psi$:

$$\psi \rightarrow e^{-i\alpha}\psi,$$

where $\alpha$ is a real constant independent of $x$, that is, having the same value everywhere and for all time. The set of all such transformations, phase change with a real constant, constitute a unitary group in one dimension, a trivial group denoted as U(1) and we refer to it as the global phase transformation. We say that the QED Lagrangian is invariant under global phase transformation. It is a "big" name for something so trivial, but the idea here is to set the language straight and distinguish this trivial case from more complicated cases yet to come when phase transformations are local, that is, dependent on $x$, rather than global.

A few words on terminology might be in order here. Phase transformation, whether global or local, are more often called "gauge" transformation. To the extent that the original definition of gauge transformation refers to the electromagnetic potential, as discussed in Chapter 6, this may be a little confusing. There is a good rationale to extend the definition of gauge transformation to include the local phase transformation and this will be explained below. Until then, we will stick to phase transformation (which is actually what it is).

Let us now consider phase transformations that are local, that is, the phase $\alpha$ is a function of $x$. Since the fields at each $x$ are considered as independent variables in the scheme of canonical quantization formalism, it is not unreasonable to consider different phase transformations at different space–time points $x$. The question now is whether or not the QED Lagrangian is invariant under such local phase transformation. It is immediately clear by observation that the QED Lagrangian is not invariant under local phase transformation; all terms in the Lagrangian except one are trivially invariant, but the "kinetic energy" term involving the differential operator is not. We have

$$\bar{\psi}e^{i\alpha(x)}(i\gamma^\mu\partial_\mu)e^{-i\alpha(x)}\psi = \bar{\psi}i\gamma^\mu\partial_\mu\psi + \bar{\psi}\gamma^\mu\psi\partial_\mu\alpha(x)$$

and the QED Lagrangian picks up an extra term $\bar{\psi}\gamma^\mu\psi\partial_\mu\alpha(x)$,

$$\mathcal{L} = \bar{\psi}(i\gamma^\mu\partial_\mu - m)\psi - \frac{1}{4}F_{\mu\nu}F^{\mu\nu} - e\bar{\psi}\gamma^\mu\psi A_\mu + \bar{\psi}\gamma^\mu\psi\partial_\mu\alpha(x).$$

At this point the gauge transformation of the electromagnetic potential $A_\mu(x)$ swings into action. As discussed in Chapter 6, $A_\mu(x)$ is determined only up to four-divergence of an arbitrary function $\Lambda(x)$, that is,

$$A^\mu + \partial^\mu\Lambda = \left(\phi + \frac{\partial\Lambda}{\partial t}, \mathbf{A} - \nabla\Lambda\right),$$

which is the original gauge transformation of electromagnetism. If we now choose the arbitrary function $\Lambda(x)$ to be equal to the local phase transformation of the Dirac field divided by the electromagnetic coupling constant $e$, that is,

$$\Lambda(x) = \frac{\alpha(x)}{e},$$

the interaction term of the Lagrangian yields another extra term that exactly cancels out the unwanted term,

$$\begin{aligned} -e\bar{\psi}\gamma^\mu\psi A_\mu &\to -e\bar{\psi}\gamma^\mu\psi\left(A_\mu + \frac{1}{e}\partial_\mu\alpha\right) \\ &= -e\bar{\psi}\gamma^\mu\psi A_\mu - \bar{\psi}\gamma^\mu\psi\partial_\mu\alpha. \end{aligned}$$

The interplay between the local phase transformation on the Dirac field and the matching choice of the electromagnetic gauge transformation "constructively conspires" to render the QED Lagrangian invariant under local phase transformations.

In sum, the QED Lagrangian

$$\mathcal{L} = \bar{\psi}(i\gamma^\mu\partial_\mu - m)\psi - \frac{1}{4}F_{\mu\nu}F^{\mu\nu} - e\bar{\psi}\gamma^\mu\psi A_\mu$$

is invariant under the local phase transformation

$$\psi \to e^{-i\alpha(x)}\psi$$

provided the gauge transformation of $A_\mu(x)$ is chosen to be

$$A_\mu(x) \to A_\mu(x) + \frac{1}{e}\partial_\mu\alpha(x).$$

With this choice, the local *phase* transformation on the Dirac field becomes interwoven with the electromagnetic gauge transformation and changes its name to local *gauge* transformation.

The freedom of gauge transformation of the electromagnetic potential thus plays an indispensable role without which the invariance under local gauge transformation cannot be upheld. This is not the first time that the electromagnetic gauge transformation is playing a critical role within the framework of the Lagrangian field theory. The quanta of electromagnetic field are, of course, massless, but if they were to have non-zero mass, say, $\kappa$, the Lagrangian would have had to contain a term $\kappa^2 A_\mu(x)A^\mu(x)$ which is clearly not invariant under the gauge transformation. The gauge invariance requires $A_\mu(x)$ to correspond to massless spin one particles, to wit, photons. Furthermore, the proof of renormalizability of QED discussed in the last chapter is also based on the gauge invariance of the Lagrangian. The roles played by gauge transformation of $A_\mu(x)$ within the framework of QED are absolutely indispensable.

The invariance of QED Lagrangian under the local gauge transformation is now to be elevated to the lofty status of a new general principle of quantum field theory, which can perhaps be extended to interactions other than electromagnetic, namely, the weak and strong interactions. To this end, we can now state the new principle, christened the gauge principle, as follows: From the way in which the freedom of gauge transformation of the electromagnetic field $A_\mu(x)$ plays the crucial role, we can define a generic field, say $B_\mu(x)$, with just such property and call it the gauge field. A gauge field is defined as a four-vector field with the freedom of gauge transformation, and it corresponds to massless particles of spin one. The gauge principle requires that the free Dirac Langrangian $\mathcal{L} = \bar{\psi}(i\gamma^\mu \partial_\mu - m)\psi$ be invariant under the local gauge transformation $\psi \to e^{-i\alpha(x)}\psi$. The invariance is upheld when we invoke a gauge field $B_\mu(x)$ such that

$$\text{(i)}\ \partial_\mu \to \partial_\mu + igB_\mu(x)$$

$$\text{(ii)}\ B_\mu(x) \to B_\mu(x) + \frac{1}{g}\partial_\mu \alpha(x)$$

where $g$ stands for the coupling constant of a particular interaction, that is, the strength of a particular force. This new gauge principle then leads to a unique interaction term of the form $g\bar{\psi}\gamma^\mu\psi B_\mu$. What the gauge principle does is that it reproduces the substitution rule as a consequence of the invariance of free Dirac Lagrangian under the local gauge transformation, thereby bypassing the classical Hamiltonian formalism for charged particles in an electromagnetic field.

Now as far as QED is concerned, however, while providing new insight and perspective, the gauge principle does not provide anything new except restating the known procedure of substitution rule. The perturbation expansion, renormalization, and renormalizability of QED work just fine without provoking such a lofty invariance requirement. The significance of this new principle lies in the fact that it provides an entirely new window through which to formulate non-electromagnetic interactions within the framework of Lagrangian field theory. Once again — for the third time in fact — the electromagnetic interaction provides a path of emulation for other interactions to follow. The most critical element in this new approach is the idea of gauge fields and this is how the gauge field theory, or more fully gauge quantum field theory of interacting particles, was born.

In the case of QED, there is one and only one gauge field, namely the electromagnetic field, that is, one and only type of photons. Photon is in a class by itself and does not come in a multiplet of other varieties. In the language of representation of a group, the photon is a singlet. Local gauge transformation involves pure numbers that are functions of $x$. The set of all such local gauge transformations form a one-dimensional trivial group U(1) which is by definition commutative. Non-commutativity will involve matrices rather than pure numbers. There is another name for being commutative called Abelian, non-commutative being non-Abelian. We refer to the local gauge transformation as Abelian U(1) transformation.

In the case of weak and strong interactions the situation becomes much more complex. The symmetries involved dictate the gauge fields to come in multiplets. In the case of weak interactions, we need three gauge fields to account for SU(2) symmetry and in strong interactions, we need eight gauge fields for SU(3) symmetry. Applying the gauge principle to these interactions is definitely more complicated and we first need to discuss the symmetry structures of basic Dirac field particles, namely, quarks and leptons, to which we now turn to in the next chapter.

# 13

# Quarks and Leptons

The elite group of particles that constitute elementary particles — the basic building blocks of all known matter in the universe — are divided into two distinct camps, a group of heavier particles called hadrons and a group of relatively lighter ones called leptons. The premier member of hadrons, proton, for example, is about 1,874 times more massive than electron, the premier member of leptons. The names "hadrons" and "leptons" originate from Greek words meaning "strong" and "small," respectively, although this distinction becomes blurred as the heaviest "lepton" turns out to be about twice as massive as proton. What really separates hadrons from leptons is more dynamical in nature than the gaps in their masses: hadrons interact via the strong nuclear force whereas leptons have nothing to do with the strong nuclear force. All particles, both hadrons and leptons interact via the weak nuclear force and electrically charged ones via the electromagnetic force.

All hadrons are considered to be composites of quarks; the baryonic sector of hadrons, that includes protons and neutrons, is considered to be composites of three quarks and the mesonic sector, that includes familiar pions, is considered to be composites of a quark and an antiquark. It has been a little over four decades now since the original quark model came into being, but despite an overwhelming indirect evidence pointing to its validity, the quark model still lacks the definitive experimental evidence in that no isolated single quark has ever been directly observed. In terms of quarks, the set of basic elementary particles reduces from hadrons and leptons to quarks and leptons, and that is where we have remained since the early 1960s.[1]

Now, quarks and leptons are all spin half particles and thus the very starting point for their description in Lagrangian field theory is the free Dirac Lagrangian density,

$$\mathcal{L} = \bar{\psi}(i\gamma^\mu \partial_\mu - m)\psi$$

where the Dirac field $\psi$ stands for each member of the quark and lepton group. In order to formulate a field theory of interacting quarks and leptons, we can now invoke the lesson gleaned from quantum electrodynamics, namely, the newly proclaimed gauge principle. As stated in the last chapter, the gauge principle demands the invariance of free Dirac Lagrangian under the local gauge transformation $\psi \to e^{-i\alpha(x)}\psi$, and the invariance is upheld by introducing a suitably defined gauge field $B_\mu(x)$ such that two simultaneous transformations are executed:

$$\text{(i)}\ \ \partial_\mu \to \partial_\mu + igB_\mu(x)$$

$$\text{(ii)}\ \ B_\mu(x) \to B_\mu(x) + \frac{1}{g}\alpha(x).$$

That an interaction is incorporated into Lagrangian this way is the very essence of gauge principle. If we now introduce a gauge field $B_\mu(x)$ and a coupling constant $g$ for each of the three interactions — electromagnetic, weak nuclear and strong nuclear — we would then have the gauge field theory for all three interactions, right? Well, not exactly... not so fast: Things would get much more complicated than that.

Over the years, we have accumulated enough data on quarks and leptons that establish definite patterns of internal symmetries, that

[1]For a readable survey of the physics of quarks and leptons, see, for example, *The Ideas of Particle Physics,* Second Edition, by G.D. Coughlan and J.E. Dodd, Cambridge University Press (1991).

is, symmetries independent of space and time, but in the mathematical spaces of internal quantum numbers such as the isotopic spin space. It turns out that quarks and leptons both exhibit doublet structures with respect to the weak nuclear force and quarks, but not leptons, harbors additional triplet structures with respect to the strong interactions. These internal symmetries define SU(2) and SU(3) symmetries, respectively, for the weak and strong nuclear interactions and they make gauge field theories a lot more complicated than the trivial U(1) symmetry of electromagnetic interaction. For one thing, whereas there is one and only one gauge field in the case of electromagnetic interaction — the photon field — the number of gauge fields needed increases to three for weak nuclear interaction and eight for strong nuclear interaction. Another reason for increased complexity — at times quite intractable — is the fact that, whereas the elements of U(1) symmetry group are pure numbers and trivially commutative, that is, Abelian, the elements of SU(2) and SU(3) symmetry groups are matrices and hence non-commutative, that is, non-Abelian. For this reason the gauge field theory for weak and strong nuclear interactions is referred to as a non-Abelian gauge field theory.

The doublet structure of quarks and leptons that defines an SU(2) symmetry for the weak nuclear force springs from their behavior with respect to weak decays, such as the well-known beta decays. To begin with, electrons and muons have their own neutrinos, called the electron-type and muon-type neutrinos and they form two pairs, similar but distinctly different from each other. When the heavy lepton, the tau, was discovered, it too was assigned its own associated neutrino, called the tau-type neutrinos. The three doublets of leptons are hence

$$\begin{pmatrix} e \\ \nu_e \end{pmatrix} \quad \begin{pmatrix} \mu \\ \nu_\mu \end{pmatrix} \quad \begin{pmatrix} \tau \\ \nu_\tau \end{pmatrix}.$$

All neutrinos are chargeless while electrons, muons and taus carry negative one unit of charges. Each doublet belongs to the fundamental representation of weak SU(2) symmetry. Taking cues from this doublet structure of leptons, the quarks are also classified in terms of three distinct doublets:

$$\begin{pmatrix} u \\ d \end{pmatrix} \quad \begin{pmatrix} c \\ s \end{pmatrix} \quad \begin{pmatrix} t \\ b \end{pmatrix}.$$

The quarks are named up ($u$), down ($d$), charm ($c$), strange ($s$), top ($t$), and bottom ($b$) quarks. The up, charm and top quarks carry charges in the fractional amount of $+2/3$, and the down, strange and bottom quarks carry charges in the amount of $-1/3$. The six leptons and six quarks, so grouped in three distinct doublets, form the basis of all known matter in the universe: quarks make up protons and neutrons that constitute atomic nuclei, the atomic nuclei forms atoms with the help of electrons swirling around them, atoms make up molecules, and so on. These twelve particles represent the basic building block's for all known matter in the universe; they and their interactions constitute what has come to be called the Standard Model of elementary particles. They are the actors of the standard model and gauge field theory is the theoretical underpinning of the model.

The masses of these quarks and leptons are not as well understood as they ought to be. The masses of charged leptons are the best known:

Mass of electron = 0.51 MeV
Mass of muon = 105.66 MeV
Mass of tau = 1777.1 MeV.

Until very recently, it has been assumed that all neutrinos are massless. The Standard Model is built on this assumption of massless neutrinos that must always travel with the speed of light. Recently, however, several experimental evidences have been uncovered that this assumption may not hold; that neutrinos may have mass, however small, and this allows conversion from one type to another (called the neutrino oscillation). According to these latest measurements (masses are inferred by the rate of one type converting into another), one can set upper limits as follows:

Mass of electron neutrino be less than 0.0000015 MeV
Mass of muon neutrino be less than 0.17 MeV
Mass of tau neutrino be less than 24 MeV.

The question of non-zero neutrino masses and the ensuing mixing of types, conversion from one-type into another, represents a serious challenge to the Standard Model that has yet to be resolved.

We referred to the doublet structures of quarks and leptons as the weak SU(2) symmetry and since there are other types of SU(2)'s in particle physics, a few words of differentiation might be in order. The oldest and most familiar SU(2) is, of course, that of the mechanical spin of particles. Then there is the SU(2) of isotopic spin of hadrons, the charge symmetry of protons and neutrons. The isotopic spin symmetry of hadrons transcends to quarks, the up and down quarks forming an isodoublet. This isotopic spin SU(2) does not, however, extend to leptons and hence is called the strong isotopic spin. The weak SU(2), is sometimes referred to as the weak isotopic spin, while similar to the strong isotopic spin as far as quarks are concerned, is a new SU(2) that encompasses both quarks and leptons.

SU($N$) is generated by $N^2-1$ generators and in the case of SU(2) the three generators are $\tau^i/2$ where $\tau^i$ $(i = 1, 2, 3)$ are the Pauli matrices. In the free Dirac Lagrangian

$$\mathcal{L} = \bar{\psi}(i\gamma^\mu \partial_\mu - m)\psi,$$

the field $\psi(x)$ now stands for a two-component field corresponding to the fundamental representation of the weak SU(2) group and the local gauge transformation $e^{-i\alpha(x)}$ is to be replaced by local weak SU(2) gauge transformation of the form

$$\exp(-i\tau^k \alpha^k(x)) \quad \text{summed over } k = 1, 2, 3.$$

Here, $\alpha^k(x)$ are three weak isotopic spin components of the local gauge and $\tau^k\alpha^k(x)$ is the required SU(2) scalar. Written out explicitly in matrix form

$$\tau^k\alpha^k(x) = \begin{pmatrix} \alpha^3(x) & \alpha^1(x) - i\alpha^2(x) \\ \alpha^1(x) + i\alpha^2(x) & -\alpha^3(x) \end{pmatrix}.$$

The simple U(1) local gauge transformation in the case of electromagnetic interaction in terms of a single function $\alpha(x)$,

$$\exp(-i\alpha(x)),$$

is now replaced, in the case of weak nuclear interaction, by SU(2) local gauge transformation in terms of three functions $\alpha^k(x)$,

$$\exp\left(-i\begin{pmatrix}\alpha^3(x) & \alpha^1(x)-i\alpha^2(x)\\ \alpha^1(x)+i\alpha^2(x) & -\alpha^3(x)\end{pmatrix}\right).$$

The local gauge transformation of the form above applies universally to leptons and quarks, both being a two-component Dirac field corresponding to the fundamental representation of weak SU(2) symmetry. The strong nuclear interactions, on the other hand, are the exclusive domains only of quarks. Leptons have nothing to do with it. Quarks carry their own signature charges, a tri-valued new attribute that has come to be called the color charges labeled red, green and blue.[2] These color charges are not related to any physically identifiable quantities, but they triple the number of quarks and provide the basis for a new color SU(3) symmetry for strong nuclear interactions; each of the quarks — $u, d, c, s, t$, and $b$ — come in three distinct varieties of red, green and blue color charges, as in red up, green up and blue up quarks. Each set of three quarks form a three-component Dirac field corresponding to the fundamental representation of the color SU(3) symmetry group. SU(3) group is generated by eight generators $\lambda^i/2$ where $\lambda^i$ $(i = 1, 2, \ldots, 8)$ are the Gell–Mann matrices. The strong color SU(3) local gauge transformation becomes

$$\exp(-i\lambda^k\alpha^k(x)) \quad \text{summed over } k = 1, 2, \ldots, 8.$$

The local gauge transformation in this case brings in eight different phase functions at a space–time point $x$. Written out explicitly in $3x3$ matrix form, suppressing the functional argument $(x)$ for the $\alpha^k(x)$ 's, we have

$$\exp\left(-i\begin{pmatrix}\alpha^3+\alpha^8/\sqrt{3} & \alpha^1-i\alpha^2 & \alpha^4-i\alpha^5\\ \alpha^1+i\alpha^2 & -\alpha^3+\alpha^8/\sqrt{3} & \alpha^6-i\alpha^7\\ \alpha^4+i\alpha^5 & \alpha^6+i\alpha^7 & -2\alpha^8/\sqrt{3}\end{pmatrix}\right).$$

[2]The need for such tri-valued new quantum numbers as well as how the ideas have evolved to what we now call color charges is described in Appendix 5, Evolution of Color Charges.

Now, a few words about the "old" SU(3) symmetry of hadrons from which this new color SU(3) symmetry is entirely different. In the early days of the quark model, the symmetry of hadrons was based on the now "old" unitary symmetry of SU(3). It was the eightfold way of the octets of mesons and octets as well as decuplets of baryons. There were only three quarks — up, down and strange — forming a triplet with respect to this "old" SU(3). This "old" SU(3) has been completely supplanted by the weak SU(2) symmetry of leptons and quarks as discussed above and the only SU(3) symmetry of the strong nuclear interaction refers to the color SU(3) symmetry.

We see that the original simple phase factor $e^{-i\alpha}$ with constant phase (now called the global gauge transformation) has come a long way under the doctrine of local gauge invariance. We have:

Abelian U(1) for the electromagnetic interaction

$$\exp(-i\alpha(x)),$$

non-Abelian SU(2) for the weak nuclear interaction

$$\exp\left(-i\begin{pmatrix} \alpha^3(x) & \alpha^1(x) - i\alpha^2(x) \\ \alpha^1(x) + i\alpha^2(x) & -\alpha^3(x) \end{pmatrix}\right),$$

and non-Abelian color SU(3) for the strong nuclear interaction

$$\exp\left(-i\begin{pmatrix} \alpha^3 + \alpha^8/\sqrt{3} & \alpha^1 - i\alpha^2 & \alpha^4 - i\alpha^5 \\ \alpha^1 + i\alpha^2 & -\alpha^3 + \alpha^8/\sqrt{3} & \alpha^6 - i\alpha^7 \\ \alpha^4 + i\alpha^5 & \alpha^6 + i\alpha^7 & -2\alpha^8/\sqrt{3} \end{pmatrix}\right).$$

The imposition of the principle of local gauge invariance, derived from the properties of QED Lagrangian, is clearly not going to be an easy task. While there is one and only one gauge field in QED, the electromagnetic field that defined the idea of gauge fields in the first place, we must now deal with multiplets of gauge fields corresponding to the regular representation of weak SU(2) and color SU(3) groups, in particular, three gauge fields for weak SU(2) and eight gauge fields for the case of color SU(3). From the matrix form of the local gauge transformation, one can readily see the non-commutativity of algebras. In the free Dirac Lagrnagian, the non-commutativity raises its head right away; the derivative terms become, using the example of weak SU(2),

$$\bar{\psi}e^{-i\tau^j\alpha^j}i\gamma^\mu\partial_\mu e^{i\tau^k\alpha^k}\psi = \bar{\psi}i\gamma^\mu\partial_\mu\psi + \bar{\psi}e^{-i\tau^j\alpha^j}[-\gamma^\mu\tau^k\partial_\mu\alpha^k]e^{i\tau^n\alpha^n}\psi.$$

In the second term, the factor $e^{i\tau^n\alpha^n}$ must be commuted through $\tau^k$ in the middle before the two phase factors can be collapsed. Clearly more complicated than in QED, the need for multiplets of gauge fields and the non-commutative (non-Abelian) algebra represent, however, only the tip of an iceberg. The imposition of local gauge principle on weak SU(2) and color SU(3) results in the Lagrangian gauge field theory that is far more complicated than anything we have seen in the evolving theories of quantum fields.

# 14

# Non-Abelian Gauge Field Theories

The non-Abelian gauge symmetry described in the last chapter is, historically speaking, a combination of new and old. The weak SU(2) and the color SU(3) symmetries of quarks and leptons are certainly "new" ideas, having been developed in the 1960s and 1970s, but the idea of a non-Abelian gauge field theory itself is an "old" one, having been proposed in 1954 by C. N. Yang and R. L. Mills. The Yang–Mills theory, as it is called, actually predates the idea of quarks by about ten years. The gauge fields of the original Yang–Mills theory had to be massless and the only known massless gauge field at that time was the electromagnetic field. The force particles then known for non-electromagnetic interactions — pions for the strong nuclear force between protons and neutrons as well as W-bosons (sometimes called the intermediate vector bosons, the IVBs) that mediated the weak nuclear force — all had mass and Yang-Mills theory remained an interesting but unrealistic idea for almost two decades. Then came the weak SU(2) and the color SU(3) symmetries of quarks and leptons and Yang–Mills formalism was accorded a powerful revival.

We can spell out the Yang-Mills formalism, the non-Abelian local gauge field theory, for a generic SU($N$) symmetry. An SU($N$) group is generated by $N^2-1$ generators $T^a$ ($a = 1, 2, \ldots, N^2-1$) satisfying the defining commutation relations

$$[T^a, T^b] = iw^{abc}T^c,$$

where $w^{abc}$ are the structure constants of the SU($N$) algebra. The Dirac field $\psi$ in the SU($N$) symmetry is an $N$-component column matrix. We demand the free Dirac Lagrangian to be invariant under SU($N$) local gauge transformation of the form

$$\psi \to \psi \exp(-iT^a\alpha^a(x)).$$

As described in the last two chapters, this invariance requires introduction of $N^2 - 1$ number of gauge fields, say $W_\mu^a(x)$, with specific prescriptions and this is where major departures from QED enter into the theory — serious complications that result from the non-Abelian nature of symmetries. The relatively "simple" recipe in the case of QED, that is,

$$\text{(i)}\ \partial_\mu \to \partial_\mu + igB_\mu(x)$$

$$\text{(ii)}\ B_\mu \to B_\mu + \frac{1}{g}\partial_\mu\alpha(x)$$

is replaced by

$$\text{(i)}\ \partial_\mu \to \partial_\mu + igT^aW_\mu^a(x)$$

$$\text{(ii)}\ W_\mu^a \to W_\mu^a + \frac{1}{g}\partial_\mu\alpha^a(x) + w^{abc}\alpha^b(x)W_\mu^c.$$

The term involving the structure constant of SU($N$) algebra is new, arising solely out of the non-Abelian properties. Another term involving the structure constant must also be introduced into the definition of gauge field tensor and this is what introduces huge complications into any non-Abelian gauge field theories. The antisymmetric electromagnetic field tensor of QED is defined as

$$F_{\mu\nu} = \partial_\mu A_\nu - \partial_\nu A_\mu,$$

but the corresponding gauge field tensor must be defined as

$$G_{\mu\nu}^a = \partial_\mu W_\nu^a - \partial_\nu W_\mu^a - gw^{abc}W_\mu^bW_\nu^c.$$

The appearance of terms containing the structure constants of SU($N$) group is what sets the Yang–Mills theory apart from the Abelian gauge theory of QED. In the case of the trivial U(1) group, there are no commutation relations and hence no structure constants. The new term in the gauge tensor field has far-reaching consequences. In contradistinction to the case of QED wherein photons do not carry electric charges and do not interact among themselves, the particles of the gauge fields in the Yang–Mills theory interact with each other, and they do so with the coupling constant $g$, the same coupling constant with which they couple to the Dirac fields, that is, quarks and leptons. This self-interaction within the gauge fields themselves is a striking departure from QED and is the signature hallmark of all non-Abelian gauge field theories. The imposition of the local gauge principle originally derived from QED to the non-Abelian symmetries results in a brand new type of interaction, the self-interactions of the gauge fields among themselves!

The non-Abelian gauge theory of color SU(3) is now a matter of transcribing the Yang–Mills formalism given above to the case of $N = 3$. The color SU(3) is generated by eight generators $T^a = \frac{1}{2}\lambda^a$ where $\lambda^a$ are the eight Gell–Mann matrices. The structure constants of SU(3) group are usually denoted by $f^{abc}$. The Dirac field $\psi$ is a three-component column matrix in the color SU(3) space, that is,

$$\psi = \begin{pmatrix} \text{red} \\ \text{green} \\ \text{blue} \end{pmatrix}$$

for each quark species, $u, d, c, s, t$ and $b$.

The eight gauge fields are called gluon fields and their quanta gluons. Gluons are to the color force, strong nuclear interaction, what photons are to the electromagnetic force, with one major difference: the coupling constant $g$ stands not only for the interaction between quarks and gluons, but also for the self-interactions among the gluons. The total Lagrangian for the non-Abelian color SU(3) symmetry is thus:

$$\mathcal{L} = -\frac{1}{4}G^a_{\mu\nu}G^{a\mu\nu} + \bar{\psi}(i\gamma^\mu D_\mu - m)\psi$$

where

$$D_\mu = \partial_\mu - ig\frac{\lambda^a}{2}W^a_\mu,$$
$$G^a_{\mu\nu} = \partial_\mu W^a_\nu - \partial_\nu W^a_\mu - gf^{abc}W^b_\mu W^c_\nu.$$

The eight gauge fields $W^a_\mu$ $(a = 1, 2, \ldots, 8)$ are the gluon fields. The quark field $\psi$ stands for a shorthand notation for space–time four-component Dirac fields that are a three-component column matrix in the color SU(3) space and the expression $\bar{\psi}(i\gamma^\mu D_\mu - m)\psi$ stands for the sum over six species of quarks, $u, d, c, s, t$ and $b$, that is,

$$\bar{\psi}(i\gamma^\mu D_\mu - m)\psi = \sum_{u,d,c,s,t,b} \bar{\psi}(i\gamma^\mu D_\mu - m)\psi.$$

This, in a nutshell, is quantum chromodynamics, QCD, the non-Abelian gauge field theory of quark–gluon interaction that is considered the origin of strong nuclear force.

Clearly, QCD is much more complicated than QED and if we could not solve the coupled equations of QED exactly, we are certainly not going to be able to find analytic solutions of QCD either. In QED, as described in previous chapters, it was possible to find approximate solutions by perturbation expansion in the electromagnetic coupling constants. The smallness of the electromagnetic coupling constant, the fine structure constant, made perturbation expansion possible. Such is, however, not the case, in general, for strong nuclear interaction. The self-interactions among the gauge fields, the gluons in this case, however, yields one very important and useful property that is shared by all non-Abelian gauge theories. It is called the asymptotic freedom. According to this asymptotic freedom, at very short distances from each other, quarks behave almost as free particles as a result of the coupling constant becoming weaker. This carves out a domain of short distances in which perturbation expansion will be valid and makes it possible to carry out some perturbative calculations in QCD. Such approach is called the perturbative QCD[1] and while some calculations of quantities of

[1]For a comprehensive treatment of the perturbative QCD, see, for example, *Foundations of Quantum Chromodynamics*, by T. Muta, World Scientific (1998).

physical interest have yielded useful results, the perturbative QCD is a subject that is still in progress. In sum, while QCD appears to be promising — and it is certainly the only field theory for strong nuclear interaction to date — the goal of constructing a successful field theory for the strong interaction is still eluding us.

The situation with respect to weak nuclear interaction is even more complicated than for the case of QCD. The group structure of SU(2) is certainly simpler than that of SU(3), but the non-Abelian gauge theory as applied to the weak nuclear force has a few grave problems all its own. Transcription of the Yang–Mills formalism to the case of weak SU(2) is, however, straightforward enough. The three generators $T^a$ are equal to $\frac{1}{2}\tau^a$, the Pauli matrices, and the structure constants are the familiar totally antisymmetric tensor $\varepsilon^{abc}$. The total Lagrangian for the non-Abelian weak SU(2) symmetry is thus:

$$\mathcal{L} = -\frac{1}{4} G^a_{\mu\nu} G^{a\mu\nu} + \bar{\psi}(i\gamma^\mu D_\mu - m)\psi$$

where

$$D_\mu = \partial_\mu - ig\frac{\tau^a}{2} W^a_\mu,$$
$$G^a_{\mu\nu} = \partial_\mu W^a_\nu - \partial_\nu W^a_\mu - g\varepsilon^{abc} W^b_\mu W^c_\nu.$$

The three gauge fields $W^a_\mu$ $(a = 1, 2, 3)$ would be the force particles for weak nuclear interactions of quarks and leptons; they would be for the weak force what photons are to the electromagnetic force. Note the use of words “would be” rather than “are” (this will be explained below). The Dirac field $\psi$ stands for a shorthand notation for the sets of three doublets of quarks and three doublets of leptons mentioned in the previous chapter, that is,

$$\begin{pmatrix} u \\ d \end{pmatrix} \quad \begin{pmatrix} c \\ s \end{pmatrix} \quad \begin{pmatrix} t \\ b \end{pmatrix}$$

for quarks and

$$\begin{pmatrix} e \\ \nu_e \end{pmatrix} \quad \begin{pmatrix} \mu \\ \nu_\mu \end{pmatrix} \quad \begin{pmatrix} \tau \\ \nu_\tau \end{pmatrix}$$

for leptons. The expression $\bar{\psi}(i\gamma^\mu D_\mu - m)\psi$ in the Lagrangian stands for the sum over six doublets.

The situation up to this point with respect to the weak nuclear interaction is in exact parallel with the case for QCD. So what is the problem? The problem is simply this. The spin one gauge particles, whether in QCD or in the case of weak SU(2), are by definition massless. In the case of gluons of QCD, this does not present a problem since gluons are deemed unobservable by the dogma of quark confinement that no color charges are to be physically observed. No one would be concerned about the mass of unobservable particles and their masses might as well be zero. In the case of weak nuclear interaction, however, the masses of the physical spin one vector mesons that mediate the weak interaction — the W-bosons — are anything but zero. They are, in fact, very massive and there is no way that these particles can correspond to the massless gauge fields $W_\mu^a$. On the other hand, if we introduce explicitly a mass term into the Lagrangian for these gauge fields, the resulting theory generates more types of infinities that cannot be renormalized. The non-renormalizability of all previous attempts to construct a field theory of weak interaction can be traced back to the presence of the mass term in the Lagrangian. The way out of this dilemma is a tortuous path called "spontaneous symmetry breaking" of local gauge symmetry. First, you adjoin the U(1) symmetry of QED to the weak SU(2) by mixing up the third component of weak gauge fields, $W_\mu^3$, with the Abelian gauge field, $B_\mu$, of electromagnetism, thus:

$$A_\mu = \cos\theta_w B_\mu + \sin\theta_w W_\mu^3$$
$$Z_\mu = -\sin\theta_w B_\mu + \cos\theta_w W_\mu^3$$

where the mixing angle $\theta_w$ is called the Weinberg angle, $A_\mu$ is identified as the physical electromagnetic potential field, and $Z_\mu$ is the newly hypothesized neutral weak boson that forms the SU(2) triplet of weak boson together with the original W-bosons,

$$W_\mu^\pm = \frac{1}{\sqrt{2}}(W_\mu^1 \pm iW_\mu^2).$$

This is U(1) × SU(2) symmetry of combined electromagnetic and weak interactions, the so-called unified theory of "electroweak" force. The story of electroweak interaction does not end here but it is only the beginning. The masses of weak bosons are very heavy and the photons, of course, must remain massless. The masses of the charged W-bosons, $W^{\pm}$, check in at about 80 times that of a proton and the neutral boson, $Z$, is about 91 times that of a proton. The idea is then to invoke a mechanism by which the weak bosons can attain non-zero masses, without explicitly bringing in the mass term into the Lagrangian that would destroy the renormalizability of the theory. This maneuver is called the mechanism of "spontaneous local symmetry breaking" or sometimes referred to as the Higgs mechanism. It is easily the fanciest maneuver in all of quantum field theory, much fancier than even the cancellation of infinities by the mass and charge renormatization.

It goes something like this. Start with the local gauge symmetric SU(2) Lagrangian as given above, with massless gauge fields, introduce a new spin zero field called the Higgs field — and hence the Higgs particle — and break the local gauge symmetry in such a way by a new interaction between the Higgs field and gauge fields so as to generate terms in the Lagrangian that look like mass terms for gauge fields. This is how the dilemma of mass problems is theoretically solved. The full name for gauge field theory for weak nuclear interaction is thus "spontaneously broken non-Abelian gauge field theory."[2] The Higgs particle that plays a crucial role in this theory has so far eluded all attempts to discover it.

The non-Abelian gauge field theories have certainly made much progress toward our understanding of strong and weak nuclear interactions. But we are nowhere near coming close to the success of Abelian gauge field theory of QED.

---

[2]For further discussions of this topic, see, for example, *Gauge Theories of Weak Interactions* by J.C. Taylor, Cambridge University Press (1976).

# Epilogue: Leaps of Faith

As stated in the opening paragraph of Prologue, relativistic quantum field theory, or quantum field theory (QFT) for short, is the theoretical edifice of the Standard Model of elementary particle physics. Looking back at its development over the past seven decades, from the early 1930s till today, one cannot help but observe a single lineage of evolution that is critically anchored in the emulation of electrodynamics. Having successfully formulated quantum electrodynamics, the theory of electrons and photons in its basic form, we have been attempting to expand the ideas of QED to other interactions within atomic nuclei, namely, the strong and weak nuclear interactions. So far, these attempts, while garnering some impressive successes, have not yet attained the same level of completeness as the QED.

As pointed out in previous chapters, Emulation of Light I, II and III in Chapters 7, 10 and 12, there are three critical stages in which the theory of electromagnetism has been emulated in order to extend the framework of QED to two nuclear interactions. It may be worthwhile to recapitulate these emulations as a way of shedding some light as to the direction of future developments. The three stages of emulation of light are basically leaps of faith. Reasonable justifications abound, but they are essentially leaps of faith.

The first leap of faith is the introduction of the concept of matter fields, as discussed in Chapter 7. The quantization of the electromagnetic field successfully incorporated photons as the quanta of that field and — this is critical — the electromagnetic field (the four-vector potential) satisfied a classical wave equation identical to the Klein–Gordon equation for zero-mass case. A classical wave equation of the $19^{\text{th}}$ century turned out to be the same as the defining wave equation of relativistic quantum mechanics of the $20^{\text{th}}$ century! This then led to the first leap of faith — the grandest emulation of radiation by matter — that all matter particles, electrons and positrons initially and now extended to all matter particles, quarks and leptons, should be considered as quanta of their own quantized fields, each to its own. The wavefunctions of the relativistic quantum mechanics morphed into classical fields. This conceptual transition from relativistic quantum mechanical wavefunctions to classical fields was the first necessary step toward quantized matter fields. Whether such emulation of radiation by matter is totally justifiable remains an open question. It will remain an open question until we successfully achieve completely satisfactory quantum field theory of matter, a goal not yet fully achieved.

The second leap of faith as far as the non-electromagnetic interactions are concerned is the way we imitated the form of interaction term in the Lagrangian, as discussed in Chapter 10. The particular, and unique, form of electromagnetic interaction that defines QED is firmly based on the Lagrangian and Hamiltonian formalism of classical mechanics. The interaction term results from the substitution rule and the latter is derived from the Hamiltonian formulation of charged particles in the electromagnetic field and that, in turn, is based on the concept of the velocity-dependent potential in the Lagrangian formalism. It is the substitution rule that gives us the interaction term, the so-called trilinear form of interaction — a Dirac field, a Dirac adjoint field and the photon field coupling at a single space–time point. This trilinear coupling is then taken to be a doctrine for all interactions and extended to strong and weak interactions as well. The trilinear coupling form for the electromagnetic interaction is derived from the substitution rule; extending it to the cases of strong and weak interactions is simply an act of emulation.

The third leap of faith comes into play when we extend the local gauge principle to non-electromagnetic interactions, as discussed in Chapter 12. Here again, the local gauge principle is a concept abstracted from the QED Lagrangian. QED itself works fine with or without the local gauge principle. In the case of QED, the local gauge principle is simply an alternative to the substitution rule. When we extend the local gauge principle to non-electromagnetic interactions, we arrive at the non-Abelian gauge field theories, quantum chromodynamics based on the color SU(3) symmetry and U(1) $\times$ SU(2) spontaneously broken local gauge field theory for electroweak interactions. They represent the current and latest stage in our formulation of quantum field theory for non-electromagnetic interactions. Extending the idea of the local gauge principle beyond QED, however, corresponds to another, the third, leap of faith.

The two of four basic forces, the electromagnetic force and gravity, are accorded well-defined and successful theoretical framework, QED and general relativity, respectively. The quest for similar successful theories for strong and weak interactions, however, has yet to achieve such lofty status. The relativistic quantum field theory for quarks and leptons can be summarized as the successful QED and its emulation for other interactions. Whether such an approach will eventually lead to theories for non-electromagnetic interactions that are as successful as QED remains an open question.

# Appendix 1: The Natural Unit System

Relativistic quantum field theory is an intricate infusion of the special theory of relativity characterized by the constant $c$, the speed of light, and quantum theory characterized by the constant $\hbar$, the Planck's constant $h$ divided by $2\pi$. It is convenient to use as the system of units consisting of these two constants plus an arbitrary unit for length, say, meters. Such a system is called the natural unit system. In terms of the standard MKS system of units, they have the values:

$$\begin{aligned} c &= 3 \times 10^{8}\,\mathrm{m/sec} \\ \hbar &= 1.06 \times 10^{-34}\,\mathrm{Joule \cdot sec\ or\ m^{2}\,kg/sec} \\ \hbar/c &= 0.35 \times 10^{-42}\,\mathrm{kg \cdot m.} \end{aligned}$$

In the natural unit system, mass and time are expressed in terms of $m^{-1}c^{-1}\hbar$ and $mc^{-1}$, respectively, where $m$ stands for meters.

It is also customary in relativistic quantum field theory to set $c = \hbar = 1$. Thus all physical quantities are expressed as powers of a length unit, say, meters. With this choice of dimensions, energy, momentum and mass become inverse lengths. The natural unit system with $c = \hbar = 1$ provides convenience to theoretical expressions since the two constants appear in virtually all formulas in relativistic quantum field theory. When a result of a theoretical calculation needs to be compared with experimental data, however, one has to reinstate the values of $c$ and $\hbar$. In the world of elementary particles, masses as well as energies and momenta are usually expressed in MeV or GeV (mega-electron-volt or giga-electron-volt) and the length in terms of fm (fermi) which is equal to $10^{-15}$ meters. For example,

$$\hbar c \approx 197\,\mathrm{MeVfm}$$
$$\frac{e^2}{4\pi} \approx 1.44\,\mathrm{MeVfm}.$$

# Appendix 2: Notation

The coordinates in a three-dimensional space are denoted by $\mathbf{r} = (x, y, z)$ or $\mathbf{x} = (x^1, x^2, x^3)$. Latin indices $i, j, k, l$ take on space values 1, 2, 3. The coordinates of an event in four-dimensional space–time are denoted by the contravariant four-vector (c and $\hbar$ are set to be equal to 1 in the natural unit system, Appendix 1)

$$x^\mu = (x^0, x^1, x^2, x^3) = (t, x, y, z).$$

The coordinates in four-dimensional space–time are often denoted, for brevity, simply by $x = (x^0, x^1, x^2, x^3)$ without any Greek indices, especially when used as arguments for functions, as in $\phi(x)$. Greek indices $\mu, \nu, \lambda, \sigma$ take on the space–time values 0, 1, 2, 3. The summation convention, according to which repeated indices are summed, is used unless otherwise specified.

The covariant four-vector $x_\mu$ is obtained by changing the sign of the space components:

$$x_\mu = (x_0, x_1, x_2, x_3) = (t, -x, -y, -z) = g_{\mu\nu} x^\nu$$

with

$$g_{\mu\nu} = \begin{pmatrix} 1 & 0 & 0 & 0 \\ 0 & -1 & 0 & 0 \\ 0 & 0 & -1 & 0 \\ 0 & 0 & 0 & -1 \end{pmatrix}$$

The contravariant and covariant derivatives are similarly defined:

$$\frac{\partial}{\partial x^\mu} = \left(\frac{\partial}{\partial t}, \nabla\right) = \partial_\mu$$

and

$$\frac{\partial}{\partial x_\mu} = \left(\frac{\partial}{\partial t}, -\nabla\right) = \partial^\mu.$$

The momentum vectors and the electromagnetic four-potential are defined by

$$p^\mu = (E, \mathbf{p})$$

and

$$A^\mu = (\phi, \mathbf{A}),$$

respectively.

# Appendix 3: Velocity-Dependent Potential

The velocity-dependent potential within the Lagrangian formalism for the case of charged particles in an electromagnetic field has far-reaching consequences in the development of quantum field theory. It is from this velocity-dependent potential that the substitution rule for the electromagnetic interaction is derived. As such it is the very foundation for the development of quantum electrodynamics, QED. The principle of local gauge invariance of the QED Lagrangian is an abstraction based on the substitution rule. The non-Abelian gauge field theories come from applying this principle of local gauge invariance to the cases of weak and strong nuclear interactions. The genesis of the non-Abelian gauge field theories, therefore, can be traced all the way back to the discovery of the velocity-dependent potential in the $19^{\text{th}}$ century. Despite such paramount importance, the subject is treated often peripherally in textbooks on classical mechanics. Here, we will briefly sketch out how the velocity-dependent potential came about.

The electric and magnetic fields in vacuum can be expressed in the form

$$\mathbf{B} = \nabla \times \mathbf{A}$$

and

$$\mathbf{E} = -\nabla\phi - \frac{\partial \mathbf{A}}{\partial t}$$

where $\mathbf{A}$ is the vector potential and $\phi$ the scalar potential ($c = 1$ in the natural unit system). The Lorentz force formula, $\mathbf{F} = q(\mathbf{E} + \mathbf{v} \times \mathbf{B})$, can then be written as

$$\mathbf{F} = q\left(-\nabla\phi - \frac{\partial \mathbf{A}}{\partial t} + \mathbf{v} \times (\nabla \times \mathbf{A})\right).$$

Using the identity

$$\mathbf{v} \times (\nabla \times \mathbf{A}) = \nabla(\mathbf{v} \cdot \mathbf{A}) - (\mathbf{v} \cdot \nabla)\mathbf{A},$$

the Lorentz force equation can be further rewritten as

$$\mathbf{F} = q\left[-\nabla\phi - \frac{\partial \mathbf{A}}{\partial t} + \nabla(\mathbf{v} \cdot \mathbf{A}) - (\mathbf{v} \cdot \nabla)\mathbf{A}\right].$$

Combining the gradient terms, we have

$$\mathbf{F} = q\left[-\nabla(\phi - \mathbf{v} \cdot \mathbf{A}) - \left(\frac{\partial \mathbf{A}}{\partial t} + (\mathbf{v} \cdot \nabla)\mathbf{A}\right)\right].$$

The vector potential $\mathbf{A}$ is a function of $x, y, z$ as well as of time $t$ and the total derivative of $\mathbf{A}$ with respect to time is

$$\frac{d\mathbf{A}}{dt} = \frac{\partial \mathbf{A}}{\partial t} + (\mathbf{v} \cdot \nabla)\mathbf{A},$$

and the force equation reduces to

$$\mathbf{F} = q\left[-\nabla(\phi - \mathbf{v} \cdot \mathbf{A}) - \frac{d\mathbf{A}}{dt}\right].$$

Now, consider the derivative of $(\phi - \mathbf{v} \cdot \mathbf{A})$ with respect to the velocity $\mathbf{v}$. Since the scalar potential is independent of velocity, we have

$$\frac{\partial}{\partial \mathbf{v}}(\phi - \mathbf{v} \cdot \mathbf{A}) = -\frac{\partial}{\partial \mathbf{v}}(\mathbf{v} \cdot \mathbf{A}) = -\mathbf{A}$$

and we have the last piece of the puzzle, namely,

$$-\frac{d\mathbf{A}}{dt} = \frac{d}{dt}\left(\frac{\partial}{\partial \mathbf{v}}(\phi - \mathbf{v} \cdot \mathbf{A})\right).$$

The Lorentz force is derivable thus from the velocity-dependent potential of the form $(\phi - \mathbf{v} \cdot \mathbf{A})$ by the Lagrangian recipe

$$\mathbf{F} = q\left[-\nabla(\phi - \mathbf{v} \cdot \mathbf{A}) + \frac{d}{dt}\frac{\partial}{\partial \mathbf{v}}(\phi - \mathbf{v} \cdot \mathbf{A})\right]$$

and this leads to the all-important expression for the Lagrangian for charged particles in an electromagnetic field

$$L = T - q\phi + q\,\mathbf{A} \cdot \mathbf{v}.$$

# Appendix 4: Fourier Decomposition of Field

The Klein–Gordon equation allows plane-wave solutions for the field $\phi(x)$ and it can be written as

$$\phi(x) = \frac{1}{(2\pi)^{3/2}} \int b(k) e^{ikx} dk$$

where $kx = k^0 x^0 - \mathbf{kr}, dk = dk^0 d\mathbf{k}$ and $b(k)$ is the Fourier transform that specifies particular weight distribution of plane-waves with different $k$'s. Substituting the plane-wave solution into the Klein–Gordon, we get

$$\int b(k)(-k^2 + m^2) e^{ikx} dk = 0$$

indicating $b(k)$ to be of the form

$$b(k) = \delta(k^2 - m^2) c(k)$$

in which $c(k)$ is arbitrary. The delta function simply states that as the solution of Klein–Gordon equation, the plane-wave solution must obey the Einstein's energy-momentum relation, $k^2 - m^2 = 0$. The integral over $dk$ therefore is not all over the $k^0 - \mathbf{k}$ four-dimensional space, but rather only over $d\mathbf{k}$ with $k^0$ restricted by the relation [for notational convenience we switch from $(k^0)^2$ to $k_0^2$]

$$k_0^2 - \mathbf{k}^2 - m^2 = 0.$$

Introducing a new notation

$$\omega_k \equiv +\sqrt{\mathbf{k}^2 + m^2} \quad \text{with only the + sign,}$$

we have $k_0^2 = \omega_k^2$ and either $k_0 = +\omega_k$ or $k_0 = -\omega_k$. Integrating out $k_0$, the plane-wave solutions decompose into "positive frequency" and "negative frequency" parts. Using the identity

$$\delta(k^2 - m^2) = \frac{1}{2\omega_k}[\delta(k_0 - \omega_k) + \delta(k_0 + \omega_k)],$$

the plane-wave solutions become

$$\phi(x) = \frac{1}{(2\pi)^{3/2}} \int \frac{d^3\mathbf{k}}{2\omega_k}(c^{(-)}(-\omega_k, \mathbf{k})e^{-i\omega_k x_0}e^{-i\mathbf{k}\mathbf{x}} \\ + c^{(+)}(\omega_k, \mathbf{k})e^{i\omega_k x_0}e^{-i\mathbf{k}\mathbf{x}}).$$

Changing $\mathbf{k}$ to $-\mathbf{k}$ in the first term, we have the decomposition

$$\phi(x) = \int d^3\mathbf{k}(a(\mathbf{k})f_k(x) + a^*(\mathbf{k})f_k^*(x))$$

where

$$f_k(x) = \frac{1}{\sqrt{(2\pi)^3 2\omega_k}}e^{-ikx} \quad \text{and} \quad f_k^*(x) = \frac{1}{\sqrt{(2\pi)^3 2\omega_k}}e^{+ikx}$$

After the decomposition into "positive frequency" and "negative frequency" parts, the notation $k_0$, as in $e^{-ikx}$, stands as a shorthand for $+\omega_k$, that is, after $k_0$ is integrated out, notation $k_0 = +\omega_k$.

# Appendix 5: Evolution of Color Charges

It has been a little over four decades since the quark model of hadrons was introduced into particle physics and during this period, and even now, the scope of its success is truly impressive. The breath and depth with which the quark model provides the basis for our understanding of hadrons and the strong nuclear interaction are absolutely undisputable. That is not to say, however, that the quark model is without a few disturbing shortcomings. From its earliest days, the quark model had to struggle with two outstanding problems, namely, that of fractional charge assignments and of what appeared to be violation of Pauli's exclusion principle.

According to the quark model, a proton is composed of two up quarks and one down quark while a neutron is made up of one up quark and two down quarks. In the units of absolute value of the electronic charge, we have

$$Q_p = 1 = 2Q_u + Q_d$$
$$Q_n = 0 = Q_u + 2Q_d$$

where $Q_p, Q_n, Q_u, Q_d$ are the charges for proton, neutron, up and down quarks, respectively. This fixes the charges for the up and down quarks to be $+2/3$ and $-1/3$, respectively. Needless to say, this is rather bizarre. Over the past four decades, we have become so accustomed to it that we accept it as a new "gospel" of physics, but the fact remains that no particles of such bizarre charges have ever been detected to date. We invoke the dogma of quark confinement that no isolated quarks should ever be observed, but such confinement has not been proven theoretically. As long as the quark confinement remains more of a prayer than a proven theory, the issue of fractional charges of quarks will remain an unerasable problem for the quark model.

The problem of quark statistics that runs into a direct conflict with Pauli's exclusion principle could have been a serious flaw of the quark model. In the same scheme that the quark contents of proton and neutron are $(u, u, d)$ and $(u, d, d)$, respectively, we have several other particles also composed of three quarks that are closely related to protons and neutrons. Of these, two particles named $N^{*++}$ and $N^{*-}$ present a serious problem with respect to Pauli's exclusion principle, one of the very basic principles of quantum physics that has never been violated to date.

The quark contents of $N^{*++}$ and $N^{*-}$ are $(u, u, u)$ and $(d, d, d)$, respectively, and according to the quark model all three up quarks in $N^{*++}$ and all three down quarks in $N^{*-}$ are completely identical to each other, respectively, in terms of all known attributes and quantum numbers. This is to say, that the $N^{*++}$ and $N^{*-}$ systems are completely symmetric with respect to interchanges among their quark constituents, a complete and direct violation of Pauli's exclusion principle which requires a system of spin half particles to be completely antisymmetric with respect to interchanges.

Ideas proposed to overcome this dilemma can be classified into two camps: in one camp, the apparent violation of the exclusion principle was to be accepted, but quarks are allowed to obey new statistics, all to its own, that up to three identical quarks can form a system in a symmetrical manner. In other words, as far as quarks are concerned, we would "change the rules." This was the approach proposed by O. W. Greenberg (University of Maryland) and is called the parastatistics for quarks. In the second camp, the idea was to "keep the rules," but invoke a new set of quantum numbers by which quarks in a symmetric three-quark system could differ from each other. A three-quark system can still be in an antisymmetric state — symmetric with respect to all then known attributes, but antisymmetric with respect to the altogether new attribute. The new attributes must necessarily have at least three different values. Keeping Pauli's exclusion principle intact by invoking an entirely new tri-valued attribute was the approach proposed by M. Y. Han (Duke University, the author of this book) and Y. Nambu (University of Chicago, now retired). This is the very origin of what has come to be called the "color" charge of quarks.

In the second approach, a new SU(3) symmetry was introduced to account for this new tri-valued attribute of quarks. The new attributes were referred to simply as new SU(3) quantum numbers and were not named in any specific way. In the original proposal by Han and Nambu, the properties of these new attributes were utilized to transform the charge assignments for quarks to the more conventional values of 1, 0 and $-1$. The original charge assignments for the up and down quarks, $+2/3$ and $-1/3$, can be shifted up by $+1/3$ to values of 1 and 0 or can be shifted downward by $-2/3$ to values of 0 and $-1$, for example.

This "shifting" however meant that the new attributes introduced to uphold Pauli's exclusion principle could be related to electric charges and hence had to be something that is physical and real, something that could eventually be detected and measured. This possibility tended to make things quite complicated for various aspects of quark physics and the idea of integer values for charges of quarks gradually fell into disfavor. Until such time as if and when isolated quarks can actually be observed and their charges measured, the idea of new SU(3) symmetry being physically related to charges seemed to be adding another layer of complexities without apparent benefit.

Several years had passed since the original proposal by Han and Nambu when a much simpler way to deal with the tri-valued attributes was put forward by M. Gell-Mann (Caltech, now retired). According to this third proposal — and this is the current basis for quark physics — the idea of a tri-valued new attribute defining a new SU(3) symmetry for quarks is good (and later fully supported by experimental data), but the new attributes are not to be related to any physically observable quantities. Insofar as quarks themselves cannot be directly observed (the dogma of quark confinement), their new attributes, the new degrees of freedom, cannot be related to anything physical either.

In this expedient abstraction, the tri-valued attributes define an SU(3) symmetry such that each species of quarks — $u, d, c, s, t$, and $b$ — comes in three distinct varieties; no physical properties are to be directly associated with the attributes but now there are not 6 but 18 different quarks. The newly differentiated three types of each species of quarks can be labeled in any set of three labels. Gell–Mann coined a new name and called the tri-valued attributes "color," as "red," "green," and "blue." Certainly a whimsical choice, but the name is as good as any other set of three labels — "1, 2, and 3," "alpha, beta, and gamma," or for that matter "vanilla, chocolate, and strawberry." All that was needed was a name with three matching labels. The name "color" stuck and the original SU(3) has since then been called the color SU(3) symmetry and the new attributes became the color charges of quarks. The color charges are to strong nuclear interaction what the electric charges are to electromagnetic interaction; they are the source charges for the strong nuclear force. In parallel to the label QED, quantum electrodynamics for electromagnetic interaction, the theory of strong nuclear interaction based on the color charges of quarks was christened QCD, quantum chromodynamics. QED and QCD are thus two of the three charter members of the Standard Model, the third one being reserved for weak nuclear interaction.

# Index

Printed in the United States
By Bookmasters